I0141216

FM 17-78

DEPARTMENT OF THE ARMY FIELD MANUAL

TANK 90-MM GUN, M47

FIELD MANUAL

By DEPARTMENT OF THE ARMY • APRIL 1955

©2013 Periscope Film LLC
All Rights Reserved
ISBN#978-1-940453-01-9
www.PeriscopeFilm.com

DISCLAIMER:

This document is a reproduction of a text first published by the Department of the Army, Washington DC. All source material contained herein has been approved for public release and unlimited distribution by an agency of the US Government. Any US Government markings in this reproduction that indicate limited distribution or classified material have been superseded by downgrading instructions promulgated by an agency of the US government after the original publication of the document No US government agency is associated with the publication of this reproduction. This manual is sold for historic research purposes only, as an entertainment. It contains obsolete information and is not intended to be used as part of an actual training program. No book can substitute for proper training by an authorized instructor.

©2013 Periscope Film LLC
All Rights Reserved
ISBN#978-1-940453-01-9
www.PeriscopeFilm.com

FIELD MANUAL⎱
No. 17–78　　⎰

DEPARTMENT OF THE ARMY
WASHINGTON 25, D. C., *15 April 1955*

TANK, 90-mm GUN, M47

		Paragraph	Page
CHAPTER 1.	INTRODUCTION	1, 2	3
2.	MATERIEL		
Section I.	General characteristics	3–4	4
II.	Gun, 90-mm, M36	5–11	8
III.	Machinegun mounts	12–17	19
IV.	Turret and armament controls and equipment	18–27	26
V.	Direct-fire sights, vision devices, and auxiliary fire-control equipment	28–37	32
VI.	Turret and gun manual and power control system	38–42	57
CHAPTER 3.	CREW DRILL, SERVICE OF THE PIECE AND STOWAGE		
Section I.	Crew composition and formations	43–45	83
II.	Crew control	46–52	84
III.	Crew drill	53–58	86
IV.	Service of the piece	59–66	93
V.	Mounted action	67–72	96
VI.	Dismounted action	73–76	104
VII.	Evacuation of wounded from tanks	77–79	109
VIII.	Inspection and maintenance	80–86	110
IX.	Destruction of equipment and stowage	87, 88	123
CHAPTER 4.	CONDUCT OF FIRE		
Section I.	Firing duties, fire commands, sensings, and moving targets	89–94	133
II.	Adjustment of fire	95–98	139
III.	Tank gunnery qualification course	99–109	148
APPENDIX I.	REFERENCES	-----	180
II.	SUBJECT SCHEDULE	-----	181
III.	NOMENCLATURE OF COMPONENTS	-----	185
Index		-----	186

*This manual supersedes TC 36, 29 December 1952; TC 18, 8 September 1953; and so much of chapter 18, FM 17–12, 3 November 1950, as pertains to Tank, 90-mm Gun, M47.

CHAPTER 1

INTRODUCTION

1. Purpose

The purpose of this manual is—

a. To give the general characteristics of the Tank, 90-mm Gun, M47.

b. To explain, in detail, the 90-mm Gun, M36, the turret controls, the fire control instruments, and the auxiliary fire-control equipment in the Tank, 90-mm Gun, M47.

c. To provide a guide for armor unit personnel in learning and teaching the fire commands, firing duties, crew drill and service of the piece for the M47 tank.

2. Scope

This manual covers materiel, crew drill, service of the piece, conduct of fire, and the tank gunnery qualification course for the Tank, 90-mm Gun, M47.

CHAPTER 2

MATERIEL

Section I. GENERAL CHARACTERISTICS

3. General

a. The basic design for the Tank, 90-mm Gun, M47, was derived from the Medium Tank, M26, which was standardized in April 1945. After 2 years of field service, a number of changes were required in order to improve the combat effectiveness of the M26 tank. These improvements included a new power plant, a torque-drive transmission with wobble-stick control for the driver, a device to clear gases from the gun tube after firing, and a cant-corrected sight mount for the gunner's telescope. The modified M26 tank was designated the M46 Patton Tank.

b. The M46 tank was accepted and standardized in 1948; but again, a number of deficiencies were discovered during field service. The development program continued; and in 1951, the Tank, 90-mm Gun, M47, made its appearance.

c. The M47 hull and suspension system are essentially the same as those on the M46; however, ballistic protection is considerably increased by the greater slope of the frontal armor. The turret design of the M47 is new, as are practically all its components, including controls, armament, and sighting equipment.

4. Description and Data, Tank, 90-mm Gun, M47

a. *Description.* The Tank, 90-mm Gun, M47 (figs. 1, 2, and 3), is a fully armored, full-track vehicle of the medium-gun tank class. It is divided into three compartments: the fighting compartment in the turret for the tank commander, gunner, and loader; the driving compartment in the front of the hull for the driver and bow gunner; and the engine compartment in the rear of the hull for the engine and transmission.

4

Figure 1. Tank, 90-mm gun, M47—left side view.

5

Figure 2. Tank, 90-mm gun, M47, with turret in traveling position.

6

Figure 3.   Tank, 90-mm gun, M47—front view.

b. Data.

Crew	Five.
Armament	One 90-mm Gun, M36; one Caliber .50 Machinegun, HB, M2; two Caliber .30 Machineguns, M1919A4.
Communication system	Radio and interphone.
Weight, fully equipped (approximately)	49 tons.
Length, overall (gun in traveling position)	23 feet 1 inch.
Length, overall (gun in firing position)	27 feet 9 inches.
Width	11 feet 6 inches.
Height, overall	10 feet 5 inches.
Ground clearance	18 inches.
Ground pressure (approximately)	13.5 pounds per square inch.
Engine	Continental, V–12, 810 hp, air-cooled.
Electrical system	24-volt.
Maximum grade-ascending ability	60 percent.
Minimum turning circle	Pivot.
Maximum fording depth	4 feet.

Section II. GUN, 90-mm, M36

5. General

a. The Gun, 90-mm, M36, is designed primarily for the medium-gun tank. It consists of four major parts: the tube, bore evacuator, counterweight, and breech mechanism.

b. The tube is formed in one piece, threaded at the breech end for attachment of the breech mechanism and at the muzzle end for assembly of the bore evacuator and attachment of the counterweight. The bore of the tube is rifled with grooves having a uniform right twist and is chrome-plated over the full length of the rifling.

c. The bore evacuator is formed by a thin-walled cylinder fitted around the forward portion of the tube to form an evacuator chamber. Eight holes, drilled into the tube and slanted at an angle of 30° toward the muzzle, connect the evacuator chamber to the bore. The bore evacuator removes residual gases from the gun tube through the muzzle end, preventing crew discomfort caused by these gases escaping into the fighting compartment.

d. The counterweight is a heavy sleeve threaded at one end for attachment to the gun tube. It is designed to balance the gun and to reduce obscuration of the target by muzzle blast.

e. The breech mechanism consists of the breech ring, breechblock with its component parts, and breech operating mechanism (figs. 4, 5, and 6).

6. Data, Gun, 90-mm, M36

Caliber	90 millimeters.
Length of bore	14 feet 9 inches.
Type of breechblock	Vertical sliding-wedge.
Maximum powder pressure	47,000 pounds per square inch.
Type of recoil mechanism	Concentric, hydrospring.
Normal recoil	12 inches.
Maximum recoil	14 inches.
Maximum elevation	+19 degrees.
Maximum depression	—10 degrees.
Weight of gun, complete	2,650 pounds.
Weight of tube only	1,750 pounds.
Ammunition and approximate velocity:	
HE and WP	2,400 feet per second.
AP	3,050 feet per second.
HVAP	4,050 feet per second.
HVAP-DS	4,100 feet per second.
HEAT	2,800 feet per second.

7. Disassembly and Assembly, Gun, 90-mm, M36

a. General. The disassembly and assembly procedures herein are intended as guides and may be altered as necessary. Because of the weight of the breechblock, extreme care must be taken to avoid injury to personnel during disassembly and assembly. At least two persons should perform the disassembly and assembly of the weapon.

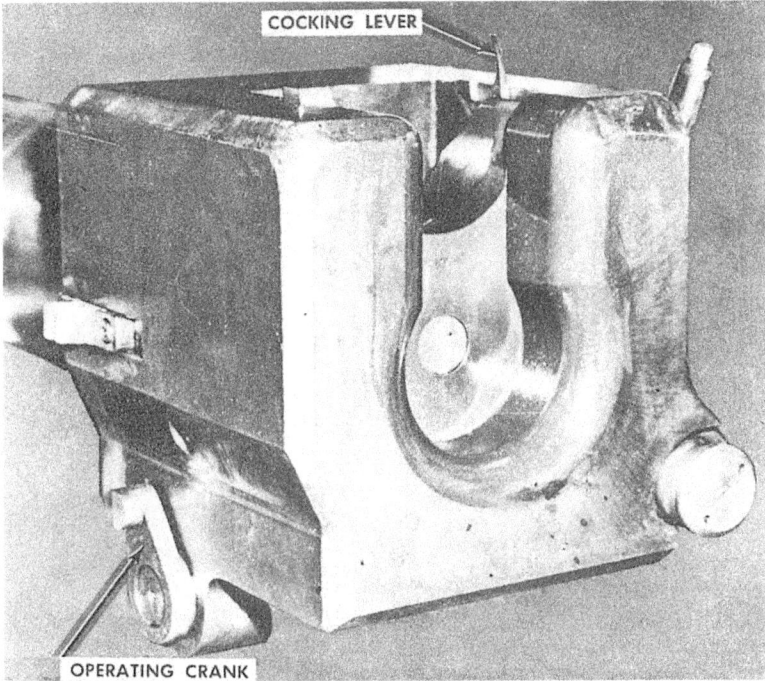

Figure 4. Breech closed—left rear view.

b. Disassembly.

(1) Open the breech slightly and inspect the chamber.

(2) Operate the firing trigger or hand firing lever to relieve the compression of the firing spring.

(3) Remove the firing spring retainer by pressing it inward and rotating it 90° until the slot in the retainer is horizontal.

(4) Remove the firing spring, hold one hand over the percussion mechanism well, and pull the cocking lever to the rear. This moves the percussion mechanism to the rear past the sear. Grasp the rear end of the percussion mechanism and remove it from the percussion mechanism well.

(5) Screw the eyebolt into the top of the breechblock and depress the muzzle of the gun fully. Attach a rope or S-hook to the eyebolt and attach it to the eye welded on the inside of the turret roof. If a rope is used, be sure there is no slack between the eyebolt and the turret eye.

> *Note.* The eye attached to the turret roof should be positioned directly above the center of the breechblock. This position is determined with both the gun and tank level. If the eye is correctly located it will prevent binding of the breechblock during disassembly

9

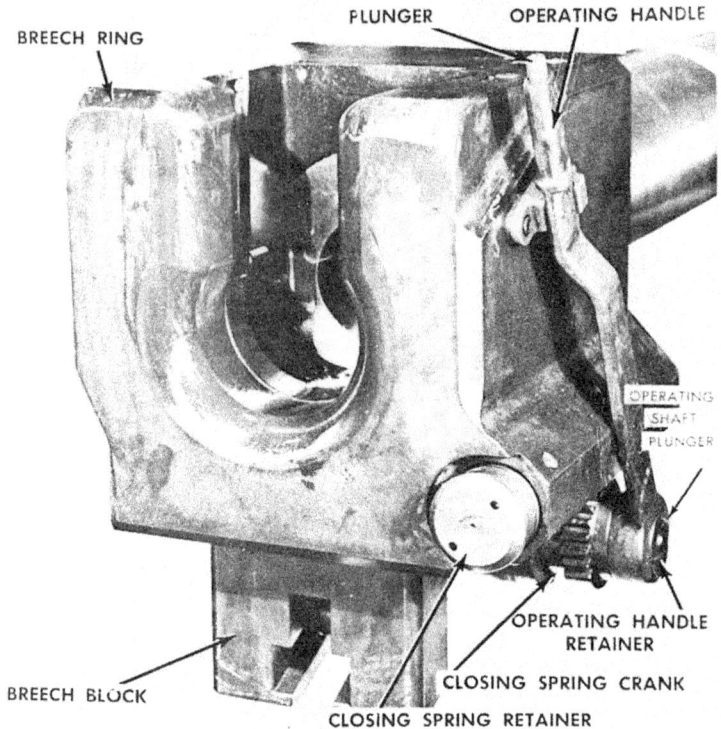

Figure 5. Breech open—right rear view.

and assembly and will prevent the bending or breaking of the breech-block eyebolts. Any repositioning of the eye should be accomplished by qualified ordnance personnel.

(6) Push to the left on the operating shaft plunger and remove the operating handle retainer. Pull the operating handle slightly to the rear, and insert the operating handle retainer into the space between the front of the closing spring piston and the gun mount. Latch the operating handle.

(7) Elevate the muzzle slightly to remove the weight of the breechblock from the operating shaft. Push the operating shaft to the left, and remove it. Catch the spacers as they are cleared by the operating shaft.

(8) Elevate the muzzle until the rear of the breechblock clears the loading notch of the breech ring. Remove the breech-block crank, taking care not to drop the crosshead rollers.

(9) Elevate the muzzle until the bottom of the breechblock is clear of the breech ring. Push the bottom of the breechblock forward until it is in line with the forward edge of the breech

10

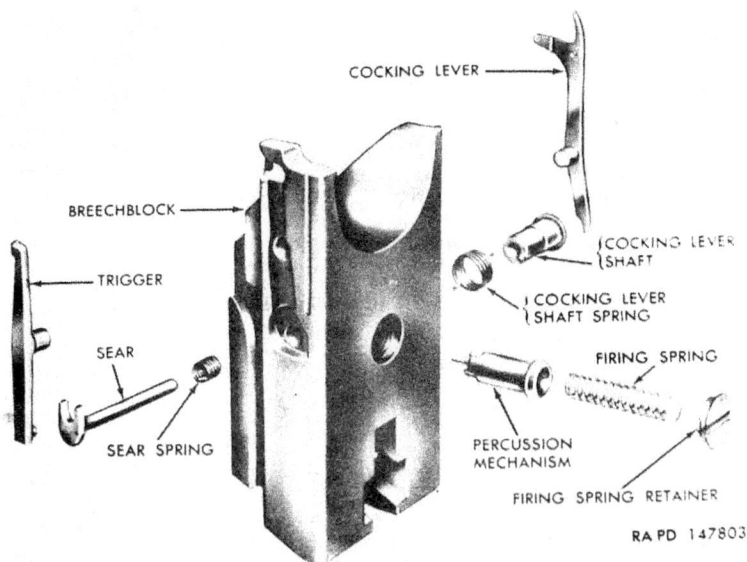

COCKING LEVER

BREECHBLOCK

TRIGGER

SEAR

SEAR SPRING

COCKING LEVER
SHAFT

COCKING LEVER
SHAFT SPRING

FIRING SPRING

PERCUSSION
MECHANISM

FIRING SPRING RETAINER

RA PD 147803

Figure 6. Breechblock and component parts—exploded view.

ring. Depress the gun, and allow the breechblock to rest flat on top of the breech ring. Be careful not to damage the quadrant seats or the firing linkage.

Caution: At no time should the hands be placed between the breechblock and breech ring during removal or replacement of the breechblock.

(10) Remove the extractors from the breech ring.

(11) Remove the cocking lever, cocking lever shaft, and cocking lever shaft spring from the right side of the breechblock.

(12) Remove the trigger, sear, and sear spring from the left side of the breechblock.

(13) It is not necessary to remove the closing spring during field disassembly. However, the closing spring may be removed using the following steps:

(a) After field disassembly, insert the operating shaft through the operating crank and closing spring crank; unlatch the operating handle, and remove the operating handle stop from the right side of the breech ring.

(b) Pull the operating handle slightly to the rear. Remove the operating handle retainer from in front of the closing spring piston, and allow the operating handle to rotate forward until the closing spring piston contacts the gun mount at the front of the closing spring housing.

11

(c) Remove the closing spring retainer screw, unscrew the closing spring retainer, and remove the closing spring.

c. *Assembly.*

(1) If the closing spring has been removed, proceed as follows:

(a) Replace closing spring, closing spring retainer, and closing spring retainer screw. Aline the closing spring retainer so that the second adjusting hole from the front will be engaged by the screw.

> *Note.* If the closing spring becomes weak, the closing spring retainer may require a different adjustment (par. 10b).

(b) Pull the operating handle far enough to the rear to place the operating handle retainer between the closing spring piston and the gun mount.

(c) Replace the operating handle stop and latch the operating handle.

(d) Remove the operating shaft.

(2) Insert the sear spring, sear, and trigger into the left side of the breechblock.

(3) Insert the cocking lever shaft spring, cocking lever shaft, and cocking lever into the right side of the breechblock. The right end of the spring fits into the hole that is nearest the cocking lever lug on the cocking lever shaft.

(4) Insert the extractors into the breech ring.

(5) Elevate the muzzle, and ease the bottom of the breechblock into alinement with the breech ring cut. Depress the muzzle, and at the same time guide the breechblock into the breech ring, making sure the extractor lips are forward. Place the breechblock crank, with the crosshead rollers, into the inclined T-slot of the breechblock through the loading notch of the breech ring.

Caution: Trip extractors from underneath the breech ring.

(6) Continue depressing the muzzle until the forward projection on the cocking lever is flush with the top of the breech ring.

(7) With the double spline of the operating shaft at 3 o'clock and the operating crank lug at 11 o'clock, insert the operating shaft through the operating crank, left spacer, breechblock crank, right spacer, closing spring crank, and hub of the operating handle. Make sure the breechblock crank stop is to the rear.

(8) Pull the operating handle to the rear, and remove the operating handle retainer from in front of the closing spring piston. Place the retainer into the hub of the operating handle after latching the handle.

(9) Remove the rope or S-hook, and unscrew the eyebolt from the breechblock.

(10) Replace the percussion mechanism, depressing the trigger plunger so the percussion mechanism can go fully forward.

(11) Replace firing spring and firing spring retainer.

(12) Cock and actuate percussion mechanism.

8. Functioning, Gun, 90-mm, M36

a. Manual Opening of Breech. Grasp the operating handle, unlatch it, and pull it to the rear and down. As the operating handle is rotated to the rear, the lug on the operating handle hub contacts a similar lug on the closing spring crank. The closing spring crank, in turn, rotates to the rear. Since it is splined to the operating shaft, it rotates the shaft, the breechblock crank, and the operating crank. The gear sector on the closing spring crank moves the closing spring piston to the rear, compressing the closing spring between the forward end of the closing spring piston and the closing spring retainer at the rear. The crosshead rollers of the breechblock crank, riding in the inclined T-slot of the breechblock, move the breechblock to the open position. As the breech opens, the trunnions of the extractors, moving in the curved extractor grooves of the breechblock, are forced forward; and the flat surfaces of the trunnions are positioned directly above the trunnion seats of the breechblock. The closing spring expands slightly, moving the breechblock upward until the trunnion seats contact the flat surfaces of the trunnions. This locks the breechblock in the open position. The extractor plunger springs expand to insure locking of the breech in the open position.

b. Cocking. When the upper arm of the cocking lever is moved to the rear by the camming surface of the breech ring, the lower arm cams the outer lug of the cocking lever shaft forward. The shaft is rotated, and the inner lug contacts the collar of the percussion mechanism and moves it to the rear. This compresses the firing spring between the firing spring stop and the firing spring retainer. The rearward movement of the percussion mechanism rotates the sear by engaging the flat portion in the center of the sear. When the collar of the percussion mechanism is far enough to the rear to clear the sear, the sear spring rotates the sear back to its normal position, causing the sear to hold the percussion mechanism in the cocked position.

c. Automatic Closing of Breech (Loading). Automatic closing of the breech occurs when a round of ammunition is loaded into the chamber. The rim of the cartridge case contacts the lips of the extractors and pushes them forward, forcing the trunnions of the extractors off the trunnion seats of the breechblock and into the curved extractor grooves. The breechblock moves upward as the expanding closing spring forces the closing spring piston to rotate the closing spring crank, operating shaft, and breechblock crank. The crosshead rollers of the breechblock crank, riding in the inclined T-slot of the breechblock, move the breechblock to the closed position.

d. Firing. As the trigger plunger is depressed, it contacts the upper arm of the trigger and pushes it to the rear. The lower arm of the trigger moves forward. Because it is in contact with the sear, it causes the sear to rotate. Rotation of the sear releases the collar of the percussion mechanism, allowing the firing spring to expand and move the percussion mechanism forward in the breechblock. As the percussion mechanism moves forward, the firing spring stop contacts the inner face of the breechblock and halts the expansion of the firing spring. The firing pin and guide continue forward under inertia. The firing pin strikes the primer of the round through the firing pin well in the forward face of the breechblock. This final forward movement compresses the retracting spring between the firing spring stop in front and the snap ring at the rear of the firing pin. After the firing pin strikes the primer, the retracting spring expands, withdrawing the firing pin from the primer and into the firing pin well.

e. Automatic Opening of Breech. When the gun recoils after firing, the lug on the operating crank forces the operating cam on the left side of the gun mount away from the gun, compressing the operating cam return spring. As the lug clears the cam, the return spring expands, moving the cam back to its normal position. During counterrecoil, the operating crank lug strikes the operating cam and is rotated to the rear. Because the operating crank is splined to the operating shaft, the operating shaft and breechblock crank rotate, moving the breechblock to the open position. At the same time, the closing spring crank moves the closing spring piston to the rear, compressing the closing spring. As the breech opens, the trunnions of the extractors, moving in the curved extractor grooves of the breechblock, are forced forward; and the flat surfaces of the trunnions are positioned directly above the trunnion seats of the breechblock. The closing spring expands slightly, moving the breechblock upward until the trunnion seats contact the flat surfaces of the trunnions. This locks the breechblock in the open position. The extractor plunger springs expand to insure locking of the breech in the open position.

f. Extraction and Ejection. As the breechblock nears the fully opened position, and after it has cleared the rear of the cartridge case, the extractor lips are rotated to the rear as the extractor trunnions move forward in the curved extractor grooves. The lips of the extractors, being in front of the cartridge case rim, extract the case from the chamber and eject it from the breech ring.

9. Recoil Mechanism, Gun, 90-mm, M36

a. General. Major components of the concentric, hydrospring-type recoil mechanism are the recoil cylinder assembly and the replenisher.

b. Recoil Cylinder Assembly.

(1) The recoil cylinder assembly is composed of the cradle, gun tube, recoil piston, and counterrecoil spring. The recoil

cylinder is formed by the inside of the cradle and the outside of the gun tube enclosed by the cradle. The recoil piston is keyed to the gun tube near the front of the cradle. The counterrecoil spring is coiled around the gun tube between the recoil piston and the rear of the cradle. When ready for operation, the recoil cylinder is completely full of hydraulic oil.

(2) When the gun is fired, it recoils or "kicks." As it moves to the rear in recoil, the oil is forced from the rear of the piston to the front of the piston; and at the same time, the counter-recoil spring is being compressed. The inside of the cradle is tapered inward from front to rear, so the clearance between the piston and cradle is greatest at the beginning of recoil. As the gun nears the end of recoil, the throttling of the flow of oil by the taper causes the gun's rearward movement to stop. When recoil is completed, the compressed counter-recoil spring starts to expand, moving the gun forward. This movement forces the oil from the front of the recoil piston to the rear of the piston. Near the end of counterrecoil, an enlarged portion of the gun tube enters the buffer chamber in the front of the recoil cylinder. The oil in the buffer chamber is displaced by the enlarged portion of the gun tube entering the chamber. Since this surface of the tube is tapered, the escape of oil from the chamber is gradually restricted, slowing down the counterrecoil action and easing the gun into battery without undue shock.

c. *Replenisher Assembly.*

(1) The replenisher assembly (fig. 7) consists of the replenisher cylinder, replenisher piston, replenisher piston spring, indicator assembly, a spring-loaded ball valve, and a hose connection. The replenisher is connected to the recoil cylinder by a flexible hose.

(2) During firing, the heat produced causes expansion of the oil. As the oil expands, it moves through the flexible hose into the replenisher and contacts the replenisher piston. The piston moves to the right, compresses the replenisher piston spring, and causes the indicator tape to wind around the screw in the indicator assembly. The indicator is a steel tape which is serrated on both edges at one end, has one serrated edge and one smooth edge in the center, and is smooth on both edges at the other end. It terminates in a smooth notch. The indicator is read as follows: both edges rough indicates the system is low on oil and should be refilled; one edge rough and one smooth indicates the proper amount of oil in the

15

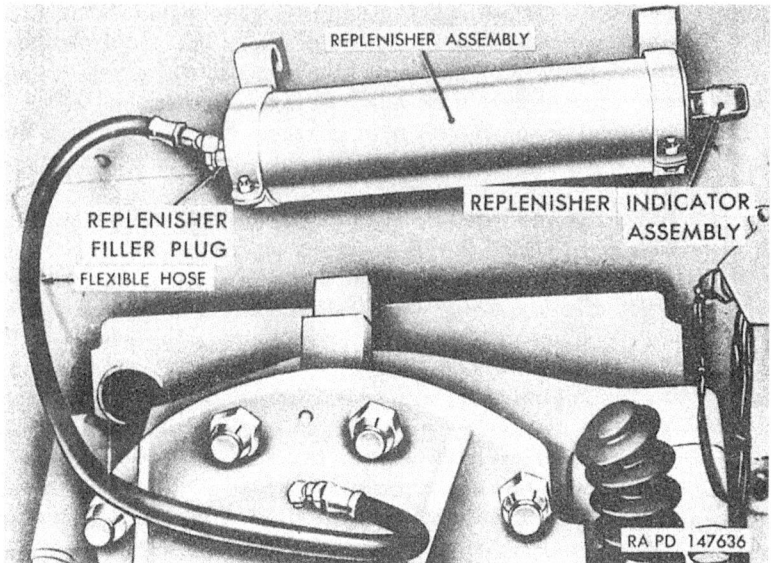

Figure 7. Replenisher assembly.

system; both edges smooth indicates an excess of oil. This excess should be removed. When the gun cools after firing, the oil contracts and the compressed replenisher piston spring expands, forcing the oil back into the recoil cylinder so the cylinder is full at all times.

d. *Checking, Filling, and Bleeding.*

(1) The recoil system is checked by use of the indicator on the replenisher (*c* (2) above). This check is performed before firing, while the recoil system is cool.

(2) If the amount of oil in the system is not correct, proceed as follows:

(*a*) If the amount of oil is insufficient, remove the filler plug from the replenisher and the nozzle from the filler gun. Fill the gun with the proper oil, screw the filler gun hose into the filler plug hole loosely, push on the plunger to force the air out, screw the hose tight, and force the oil into the replenisher. Repeat until the proper amount of oil is indicated (by one rough and one smooth edge on the indicator tape).

(*b*) If there is too much oil in the system, remove the filler plug, hold a rag under the filler plug hole, and drain the excess oil from the replenisher onto the rag. To drain the excess oil, it is necessary to push in on the spring-loaded ball valve in the filler plug hole with the nozzle of the filler gun.

Note. Use gradual pressure on the ball valve to control the amount of oil flowing from the replenisher. Check the indicator tape frequently to avoid draining too much oil.

(3) During firing, the system should be checked periodically for leaks and excessive oil due to overheating. After firing for a period of time, the indicator tape may show both edges smooth as a result of the recoil oil expanding. It is not necessary to drain oil from the replenisher when this occurs. However, if the long notches are exposed, the crew should drain enough oil so that the two smooth edges are exposed.

e. Bleeding. If the recoil of the gun is excessive and the replenisher shows the correct amount of oil, there may be air in the recoil system. The recoil system can be bled by the turret artillery mechanic.

10. Malfunctions, Gun, 90-mm, M36

a. General. A malfunction is an unintentional cessation of fire caused by a failure of some part of the gun, the mount, or the ammunition. Malfunctions are divided into four general classes: failure to load, premature firing, failure to fire, and failure to extract and eject. The malfunctions discussed here are the most common ones but do not necessarily include all malfunctions which may occur.

b. Failure to Load—Breech Will Not Fully Close.

Cause	Correction
Dirty, bulged, or dented round of ammunition.	Remove round and attempt to load another.
Dirty chamber	Remove round, clean chamber, and attempt to load.
Gun out of battery (obstruction between gun and mount or too much recoil oil).	Remove obstruction or drain excess oil from replenisher.
Improper adjustment of closing spring	Remove closing spring retainer screw, tighten retainer until the next hole in the retainer is in line with the screw hole, and replace the screw.

c. Premature Firing—Gun Fires as Breech Closes.

Cause	Correction
Jar of breechblock closing causes sear to release percussion mechanism because of weak sear spring.	Replace defective part.

d. Failure to Fire.

Cause	Correction
Dirt, wax, or excess grease in percussion mechanism well, preventing free movement of percussion mechanism.	Clean percussion mechanism and well.
Worn or broken cocking lever, cocking lever shaft, or cocking lever shaft spring.	Replace defective part.
Worn or broken firing pin, firing spring, sear, or sear spring.	Replace defective part.
Defective primer in the round of ammunition.	Remove round, load another round, and fire.

e. Failure to Extract and Eject.

Cause	Correction
Weak, rusty, or grease-packed operating cam return spring causes breech to fail to open on counterrecoil.	Replace or clean defective part.
Broken or missing extractor plungers cause extractor trunnions to slip off the trunnion seats of breechblock, allowing breechblock to jam the cartridge case in the breech ring.	Replace missing or defective part.
Defective rim of cartridge case causes extractor lips to pull through.	Pry or ram case out of chamber.

11. Care, Cleaning, and Lubrication, Gun, 90-mm, M36

a. General.

(1) It is of vital importance to keep all materiel in proper condition for immediate service. Tools, accessories, and cleaning, lubricating, and preserving materials are provided and must be used to keep the materiel in proper operating condition.

(2) Proper lubrication with proper lubricants at specified intervals is essential to the care and preservation of the materiel.

(3) All protective covers for the gun and mount should be installed when the gun is not in service. If the materiel is not to be used for a considerable length of time, all exposed, unpainted surfaces should be cleaned with rifle bore cleaner, dried thoroughly, and covered with a coat of rust preventive compound.

(4) During disassembly and assembly, clean disassembled parts thoroughly before oiling and assembly. A steel hammer must not be used directly on any part. If a copper or lead hammer is not available, use a wood block as a buffer. Always use the tool intended for the part. Tools and accessories, as well as the materiel, should be kept clean, free from rust, and protected with preservative oil.

(5) The quadrant seats on the breech ring must be protected. Do not place tools or other articles upon them which might cause scoring.

b. Before Firing.

(1) *Tube.* Check bore and chamber for dirt and obstructions. Clean as necessary, but do not lubricate.

(2) *Breech mechanism.* Examine the breech mechanism for functioning and freedom from corrosion. Clean and lubricate.

c. During Firing. Be alert for any malfunction. Lubricate as necessary.

d. After Firing.

(1) *Tube.* After firing and for the next 2 days, clean the bore thoroughly with rifle bore cleaner, making sure that all

surfaces in the bore and chamber are well coated. Wipe the tube dry each time before applying additional bore cleaner. On the third day after firing, clean the bore with bore cleaner. If the gun will be fired in the next 24 hours, do not wipe the tube dry. If the gun is not to be fired, dry the tube and coat it with the prescribed oil.

(2) *Breech mechanism.* Disassemble, clean with bore cleaner, lubricate, and assemble. Check for proper functioning and condition of parts.

(3) *Bore evacuator.* The bore evacuator should be cleaned each time the gun is cleaned. More frequent cleaning may be necessary if the residual gases from the firing of ammunition are not cleared from the gun tube by the bore evacuator. To clean, remove the counterweight and the evacuator can. Clean the jets by inserting a piece of wire through each of them. Clean the exterior of the tube and the interior of the evacuator can with bore cleaner. Coat the cleaned parts with the prescribed preservative, and reassemble them.

Caution: Be careful not to damage the sealing lips while removing and replacing the bore evacuator.

Section III. MACHINEGUN MOUNTS

12. General

a. The firepower of the caliber .50 and caliber .30 machineguns greatly increases the shock effect of the M47 tank.

b. This section describes and illustrates the mounts for the Machinegun, Caliber .50, M2, and Machinegun, Caliber .30, M1919A4, and furnishes essential information for the crew to mount and operate these weapons.

13. Combination Gun Mount, M78

a. General. The combination Gun Mount, M78, will mount a caliber .30 machinegun coaxially with the 90-mm gun (fig. 8). The machinegun mount is secured to the tank-gun cradle and is part of the tank-gun mount. Therefore, the tank-gun mount is called the combination gun mount. This mount is held and pivoted within the turret on two trunnions. When the main armament is elevated and traversed, the machinegun is moved correspondingly, thus making the movement of the machinegun coaxial with that of the main armament.

b. Machinegun Mount. In the M47 tank, the coaxial machinegun mount is located on the left side of the M78 combination gun mount. Two bracket assemblies, front and rear, hold the machinegun when mounted (fig. 9).

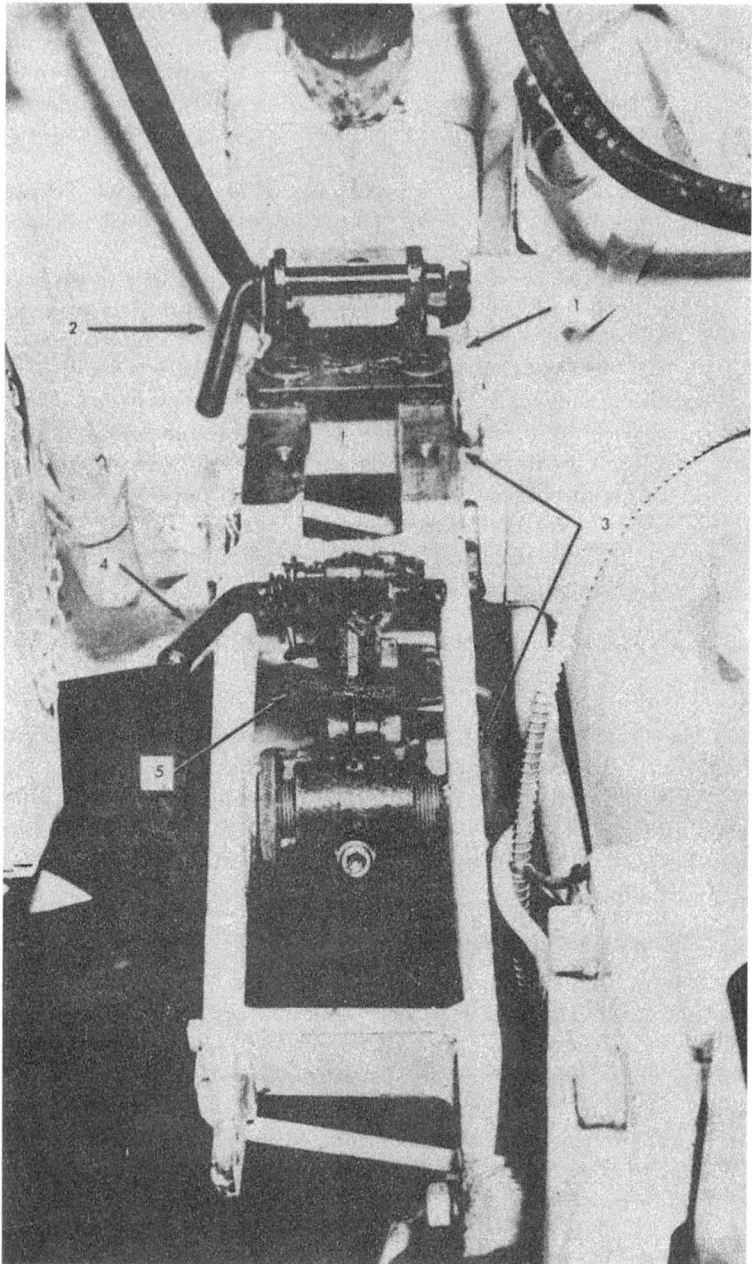

1	Front mounting bracket	4	Machinegun rear locking pin
2	Machinegun front locking pin	5	Elevating and traversing mech-
3	Machinegun cradle		anism

Figure 8. Caliber .30 machinegun mount.

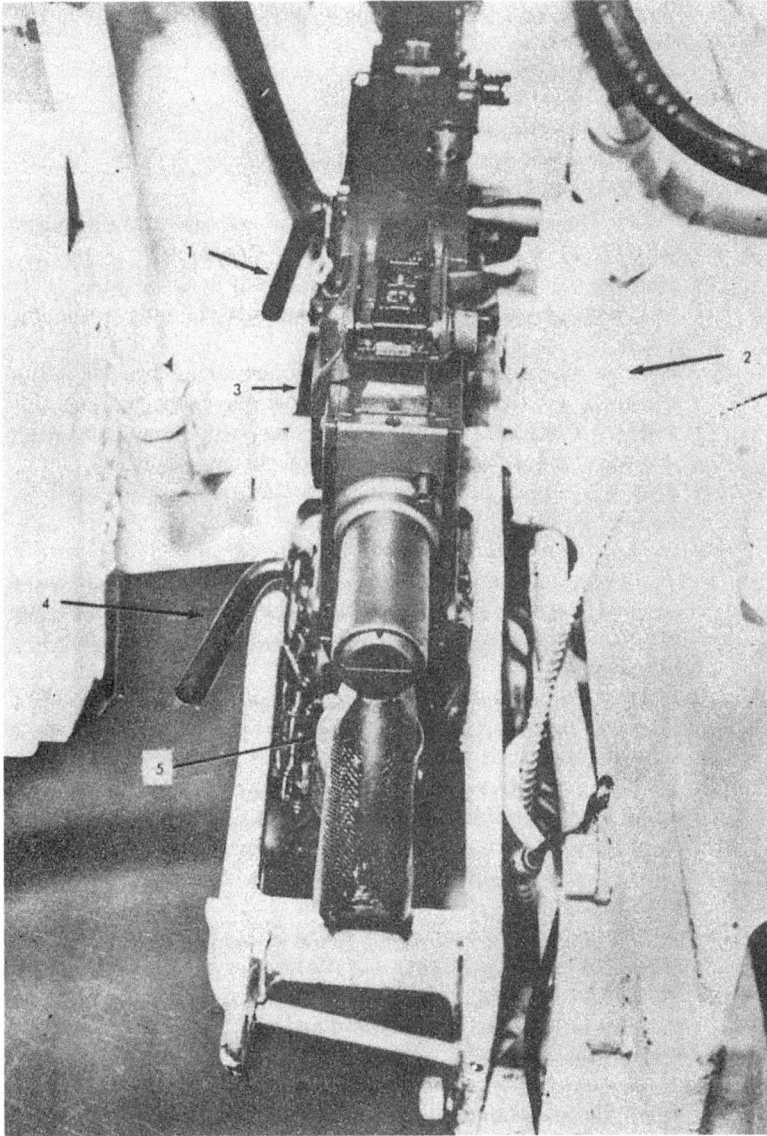

1 Machinegun front locking pin   4 Machinegun rear locking pin
2 Combination gun mount, M47   5 Elevating and traversing mechanism
3 Coaxial machinegun

Figure 9. Caliber .30 machinegun, installed.

c. Component Parts. The component parts of the Combination Gun Mount, M78, are—
   (1) Front mounting bracket.
   (2) Machinegun front locking pin.
   (3) Coaxial machinegun cradle.
   (4) Elevating and traversing mechanism.
   (5) Machinegun rear locking pin.
   Note: Check bolts on front mounting bracket frequently for tightness.

d. Installation of Machinegun, Caliber .30, M1919A4.
   (1) Pull out the machinegun front and rear locking pins.
   (2) Insert the muzzle of the machinegun into the hole in the gun shield.
   (3) Set the machinegun into the front mounting bracket, aline the front mounting holes, and insert the front locking pin.
   (4) Pull up the elevating and traversing mechanism, and aline it with the rear mounting holes of the machinegun.
   (5) Insert the rear locking pin, and the machinegune will be secured in the coaxial gun mount.

e. Adjustment of Firing Solenoid (fig. 10).
   (1) The adjustment screw is located at the bottom of the solenoid support, between the solenoid and the machinegun elevating bracket. The setscrew is located on the machinegun elevating bracket.
   (2) Cock the machinegun, and loosen the two jamnuts with a $\frac{7}{16}$-inch open-end wrench. Loosen the setscrew with a screwdriver.
   (3) Using a screwdriver, turn the adjustment screw clockwise to move the solenoid plunger away from the trigger of the machinegun; turn the adjustment screw counterclockwise to move the solenoid plunger closer to the trigger of the machinegun.
   (4) Adjustment is correct when there is $\frac{1}{32}$-inch clearance between the solenoid plunger and the trigger of the machinegun.
   (5) When the adjustment is completed, tighten the setscrew and the two jamnuts on the solenoid adjusting screw. Check for final adjustment.

f. Loading and firing Coaxial Machinegun, Caliber .30, M1919A4.
   (1) Load the magazine (fig. 11), and run the belt through the feedway and into the machinegun.
   (2) To halfload the machinegun, pull the bolt handle to the rear and release it. Repeat the operation to full-load.
   (3) To fire electrically—
      (*a*) Turn on the caliber .30 machinegun switch.
      (*b*) Press either of the gunner's firing triggers or the commander's firing trigger to achieve the desired burst.

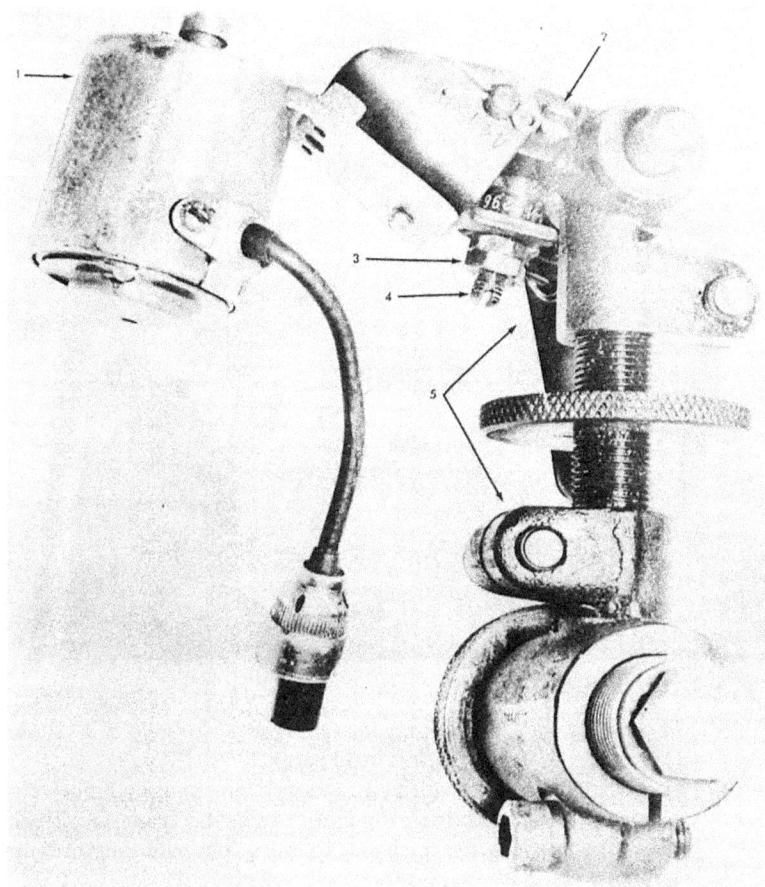

1 Firing solenoid
2 Setscrew
3 Jamnuts
4 Solenoid adjusting screw
5 Elevating and traversing mechanism

Figure 10. Firing solenoid, coaxial machinegun.

g. Boresighting Coaxial machinegun (fig. 12). The coaxial machinegun must be alined with the main armament. The back plate and bolt groups must be removed to boresight.

    (1) To adjust the machinegun for elevation—

        (*a*) Loosen the two upper socket head screws.

        (*b*) Turn the elevating adustment screw clockwise to elevate the muzzle of the machinegun; turn the elevating screw counterclockwise to depress the muzzle of the machinegun.

        (*c*) When the machinegun is sighted on the target, secure the two socket-head screws.

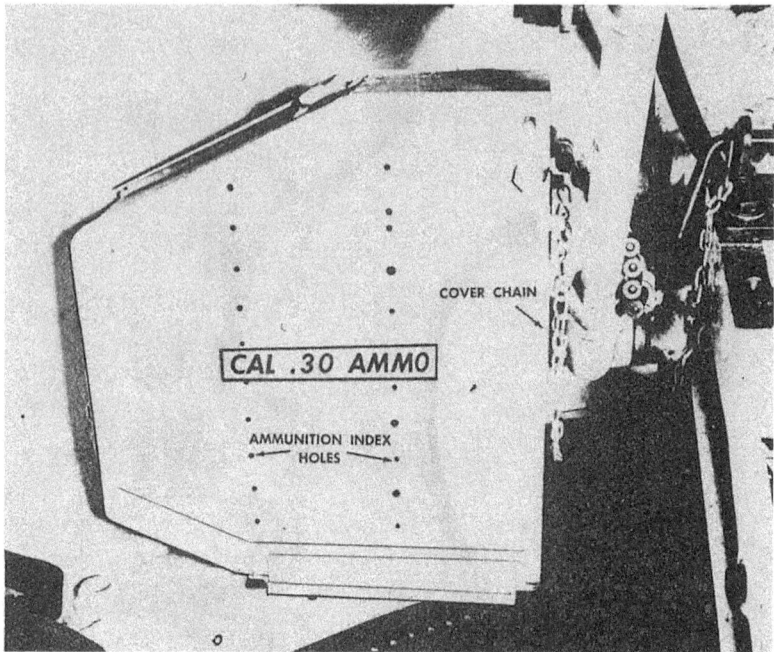

Figure 11. Magazine, caliber .30 coaxial machinegun, 750-round capacity.

   (2) To adjust the machinegun for traverse—

     (a) Loosen the lower socket head screw.

     (b) Turn the traverse adjustment screw clockwise to move the machinegun muzzle to the right; turn the traverse adjustment screw counterclockwise to move the machinegun muzzle to the left.

     (c) Tighten the lower socket head screw.

  h. Removal of Machinegun, Caliber .30, M1919A4.

   (1) Clear the machinegun.

   (2) Remove the front and rear locking pins, and lift out the machinegun.

14. Caliber .50 Machinegun, M2, Turret-Mounted

  a. General. The Machinegun, Caliber .50, M2, is located on the turret roof. The turret-mounted gun is used against both ground and air targets and is controlled and fired manually. The machinegun and the cradle assembly may be removed as a unit. When the machinegun cradle is removed, replace the cover plug assembly. When the tank is in motion, the machinegun may be locked in place by the machinegun traveling lock located forward of the commander's cupola.

  b. Pintle Mount. The pintle mount consists of—

   (1) Pintle stand assembly.

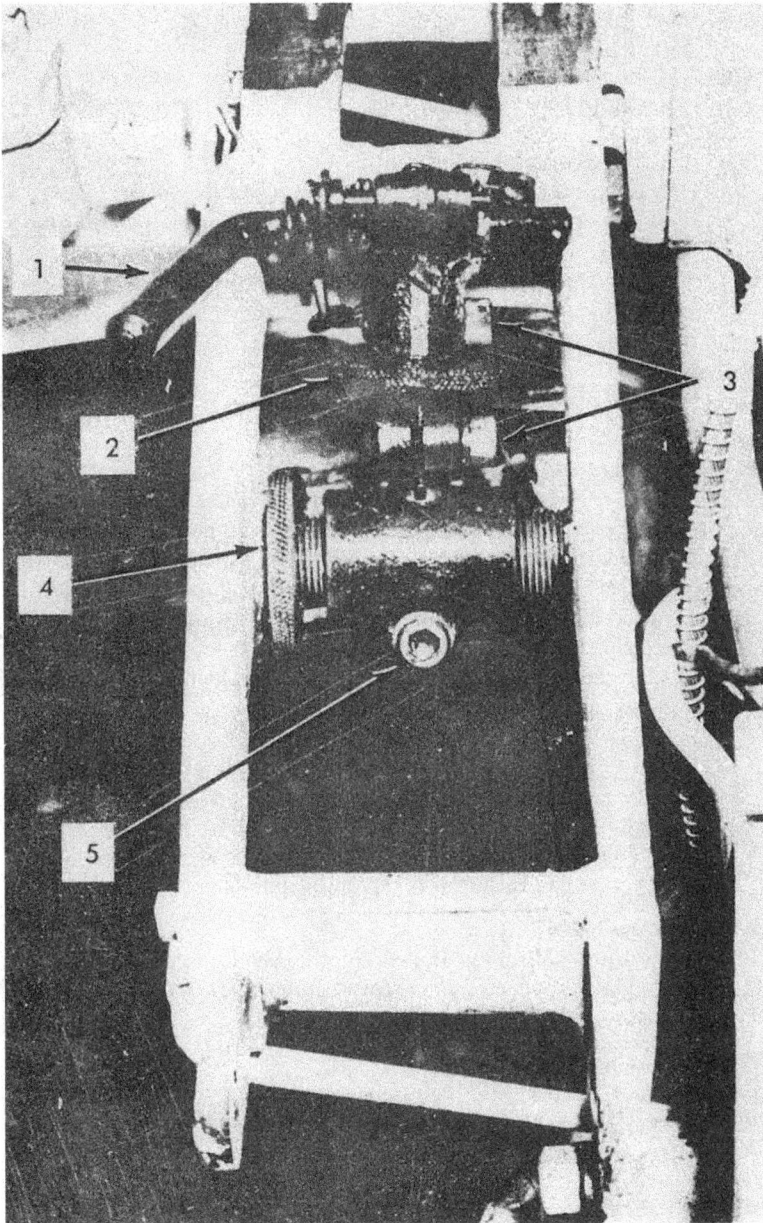

1 Machinegun front locking pin   4 Traversing adjustment screw
2 Elevation adjustment screws   5 Lower socket head cap screw.
3 Upper socket head cap screws

Figure 12. Elevating and traversing mechanism, caliber .30 coaxial machinegun.

    (2) Lock handle.

    (3) Cover plug assembly.

    (4) Cradle assembly.

    (5) Front and rear locking pins.

  c. Installation.

    (1) Remove cover plug assembly.

    (2) Place pintle lock in open position (handle down).

    (3) Insert cradle assembly in pintle stand, and lock it (handle up).

    (4) Aline front and rear machinegun mounting holes with holes on cradle assembly.

    (5) Insert front and rear locking pins.

  d. Removal.

    (1) Clear the machinegun.

    (2) Reverse the procedure for installation.

15. Caliber .30 Machinegun, Flexible Ball Mount

  a. General. A Machinegun, Caliber .30, M1919A4, is mounted in the right side of the driver's compartment in the flexible ball mount and is referred to as the bow gun. The bow gun is controlled and fired manually. It is capable of +24-degrees elevation and −10-degrees depression. It is primarily an antipersonnel weapon.

  b. Installation.

    (1) Aline front and rear mounting holes of the machinegun with holes on the mount.

    (2) Insert front locking pin.

    (3) Swing bow gun locking rod out from side of hull.

    (4) Insert locking rod in machinegun rear locking hole.

  c. Removal.

    (1) Clear the machinegun.

    (2) Reverse the procedure for installation.

16. Tripod Mount, M2

A tripod Mount, M2, is provided for the Machinegun, Caliber .30, M1919A4. The tripod mount is stored in the left front fender box when not in use.

17. Maintenance, Gun Mounts

Maintenance on the M78 combination gun mount, pintle mount, and flexible ball mount, by the using arm, is limited to inspection and normal care, cleaning, and lubrication.

Section IV. TURRET AND ARMAMENT CONTROLS AND EQUIPMENT

18. General

This section describes, locates, and illustrates the various turret controls and furnishes necessary information for the proper operation of the turret on the M47 tank.

19. Gunner's Power Control

a. The gunner's power control handle (fig. 13) is located directly in front of the gunner's position. It provides an integrated system of power control for the turret and tank gun and enables the gunner to traverse the turret and elevate or depress the gun by moving the one control handle.

1 Gunner's power control handle	4 Manual elevation control handle
2 Turret power control box	5 Manual traverse control handle
3 Hand grip (dummy)	

Figure 13. Gunner's turret controls.

b. The turret can be traversed 360° in either direction in the following manner: To traverse right, rotate the control handle clockwise; to traverse left, rotate the control handle counterclockwise. The distance the handle is rotated from neutral determines the speed of turret traverse. The maximum speed of turret traverse is four complete revolutions per minute.

c. The gun can be elevated to 19° above horizontal and depressed to 10° below horizontal at any rate up to 4° per second. To elevate, push the bottom of the control handle away from the gunner; to depress, pull the bottom of the control handle toward the gunner. The distance the handle is moved from the neutral position determines the speed of elevation or depression.

20. Tank Commander's Power Control

a. The tank commander has the same power control of the turret and gun that the gunner has, but the commander's control handle is equipped with an override lever. The control handle (fig. 14) is located on the turret wall to the tank commander's right front.

b. The commander can take the power control of the turret from the gunner by pressing the override lever on the control handle. To elevate the gun, press the override lever and push the bottom of the commander's control handle away from the commander's position. To depress the gun, press the override lever and pull the bottom of the control handle toward the commander's position. To traverse the

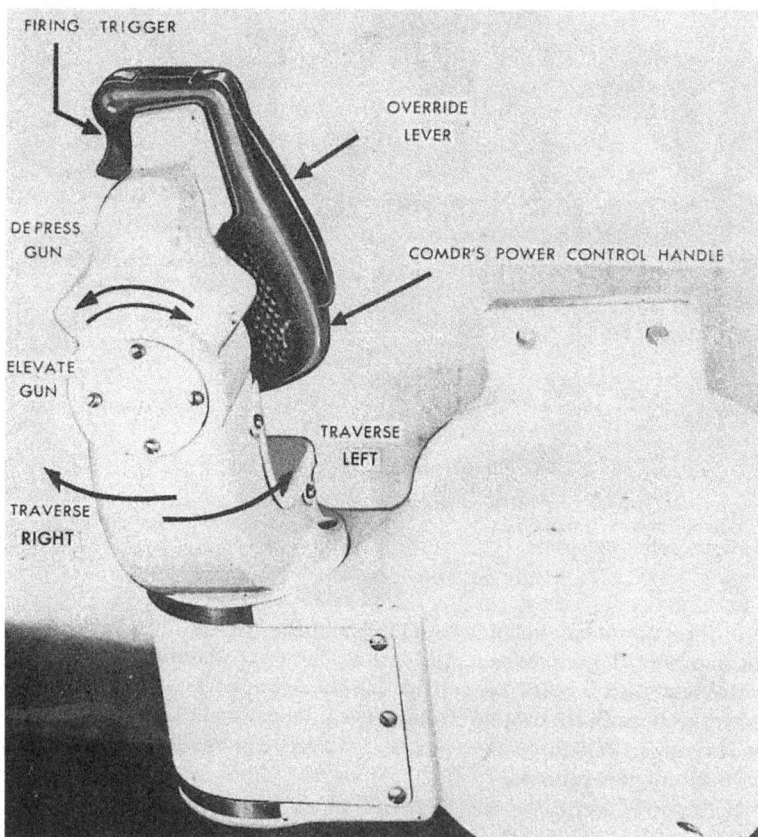

Figure 14. Commander's power control handle.

28

turret, press the override lever and rotate the power control handle in the desired direction of traverse. The distance the handle is moved from neutral determines the speed of traverse, elevation, and depression. The gun can be elevated or depressed while the turret is being traversed.

21. Gunner's Manual Controls

a. *Manual Traverse Control Handle.* The gunner's manual traverse control handle (fig. 13) is located above and to the right of the gunner's seat. To traverse the turret manually, grasp the manual traverse control handle, squeeze the release lever on the handle, and rotate the handle in the desired direction. The speed of traverse is regulated by the rate of handle rotation. A no-back mechanism automatically holds the turret in position and prevents turret drift when the vehicle is not in a horizontal position. The gunner can traverse the turret manually while the turret motor is on.

b. *Manual Elevation Control Handle.* The gunner's manual elevation control handle is located to the left of the gunner's power control handle (fig. 13). It is connected to a hydraulic pump which, by directing a flow of oil, causes the gun to be elevated or depressed. To elevate the gun, rotate the handle clockwise. To depress the gun, rotate the handle counterclockwise. The gun cannot be elevated or depressed manually while the turret motor is on.

c. *Accumulator Hand Pump and Handle.* A hand pump is located to the right of the gunner. This pump is used to charge the manual elevation system. The pump is operated by means of a handle located to the right of the gunner. To charge the system, move the handle up and down until the gun responds quickly to movements of the manual elevation control handle.

22. Loader's Traverse Safety

a. The loader does not have controls which will permit him to traverse the turret. However, for his safety while he is stowing ammunition or removing ammunition from the stowage compartment, a loader's traverse safety and indicator light are installed on the left wall of the turret.

b. When the traverse safety switch is turned to the ON position, the indicator light will glow and the turret may be traversed in power by the gunner or commander. When the switch is in the OFF position, the turret will not traverse in power. The loader should never attempt to remove ammunition from beneath the turret floor when the indicator light is on.

23. Turret Power Control Box

a. The power control box (fig. 13) is directly in front of the gunner's position. On it are mounted indicator lights and switches that control the turret motor and the electrical firing circuits.

b. The turret motor switch is on the right side of the top of the power control box. This switch must be in the ON position before the turret can be operated in power. When the switch is on, the indicator light below it will be on. When the turret motor is running, it will be noted that the noise level is quite high. This is because of an inherent characteristic of the system, and is not cause for alarm if the system has been properly serviced and maintained.

c. The 90-mm gun switch is in the center of the top of the power control box. This switch controls the electrical firing circuit for the main armament and must be in the ON position before the 90-mm gun can be fired electrically. When the 90-mm switch is on, the indicator light below it will be on.

d. The coaxial machinegun switch is on the left side of the top of the power control box. This switch controls the electrical firing circuit for the coaxial machinegun. The coaxial machinegun switch must be on before the coaxial machinegun can be fired electrically. When the machinegun switch is on, the indicator light below it will be on.

24. Firing Controls

a. The gunner is provided with firing triggers, a hand firing lever, and gun switches to control the firing of the 90-mm gun and the coaxial machinegun. Firing triggers which permit electrical firing by the gunner are located as follows: one on the front of the manual elevation control handle, one on the front of the gunner's power control handle, and one on the front of the hand grip to the left of the power control handle. The hand firing lever is located on the upper right side of the gun cradle and is easily accessible from the gunner's position. This lever permits manual firing of the 90-mm gun.

b. The tank commander may fire the 90-mm gun or the coaxial machinegun by using the firing trigger on the front of the tank commander's power control handle. The firing trigger for the tank commander provides electrical firing only. It will not function unless the necessary switches are on and the commander has taken power control of the turret by pressing the override lever on his power control handle.

c. The loader's reset safety, on the left front of the turret wall, operates to open and close the electrical firing circuit for the 90-mm gun. The safety is reset by pushing in the button. The loader must perform this operation after each round is fired. When the circuit is closed and ready for firing, the indicator light on the reset safety box will be on. The later M47 tanks do not have the loader's reset safety.

d. The 90-mm manual safety is on the top side of the gun mount. When the safety is to the rear, the gun can be fired. When the safety is forward, the firing plunger is inoperative, and gun cannot be fired.

25. Master Relay Switch

The master relay switch, located on the driver's instrument panel, is the control switch for all the electrical power in the tank. This switch must be in the ON position before the turret can be operated in power, before the guns can be fired electrically, or before the radio or interphone system will function.

26. Locks

a. *Turret Lock.*

(1) The turret lock, located to the right of the gunner's seat, holds the turret stationary by means of a gear segment which engages the turret ring teeth. The turret lock should be in the locked position while the vehicle is in motion, unless the turret is to be traversed.

(2) To unlock the turret, turn the turret lock handle clockwise until the gear segment disengages from the turret ring gear.

(3) To lock the turret, turn the handle counterclockwise. Do not use power traverse to check the turret lock.

b. *Gun Traveling Lock.*

(1) The gun traveling lock, located on the rear deck of the tank, is used to keep the gun in the locked position to avoid excessive wear of the traversing and elevating mechanism caused by vibration while the tank is moving.

(2) To open the lock, lift the lever from the top of the cap and unscrew the lever bolt from its bracket by turning the lever counterclockwise. Swing the cap and lever back. Lay the traveling lock flat on the top deck. Before the traveling lock can be raised to the vertical position again, the stop pawl must be released. To lock, swing the cap over the gun tube and tighten it by turning the lever clockwise. Rotate the lever into position over the cap, and push the lever down.

27. Putting Turret Into Power Operation

The following steps are performed in putting the turret into power operation. (See paragraph 68 for individual crew duties.) Remember the word "ACUTE" for a key to the steps.

A—*Alert crew*_____ Insure that crew is in safe position; check the immediate area for obstructions.

C—*Check oil*_____ Oil in hydraulic oil reservoir should be at FULL mark on bayonet gage.

U—*Unlock turret*_____ Traverse manually to make sure the turret is unlocked. Return the manual traverse control handle to the latched position over the dump valve micro switch.

T—*Turn on power*_____ Power control handles should be in the neutral position and the loader's traverse safety on.

E—*Elevate and traverse*_____ Operate in power to make sure the controls are functioning properly.

Note. The main or auxiliary engine should be running and charging the tank batteries while the turret is in power operation.

Section V. DIRECT-FIRE SIGHTS, VISION DEVICES, AND AUXILIARY FIRE-CONTROL EQUIPMENT

28. General

The Tank, 90-mm Gun, M47, is equipped with primary and secondary means for direct fire, necessary equipment for indirect laying of the tank gun, and the vision devices required by the crew for operation of the tank with all hatches closed. This section contains information concerning location, adjustment, and use of these various pieces of equipment.

29. Range Finder, M12; Superelevation Transmitter, M22; and Inverter (7633698)

a. The Range Finder, M12 (fig. 15) (see appendix III), is the primary direct-fire sight in the M47 tank. The range finder is operated by the gunner, who indexes the proper ammunition, ranges on the target, tracks the target if necessary, and fires the weapons. The binocular eyepieces of the range finder are stationary with respect to the movable section of the instrument. This results in the eyepieces remaining in a fixed position within the turret as the gun and range finder are elevated or depressed.

b. The gunner uses the range finder with the scales and gun-laying reticle viewed through the left eyepiece; the stereoscopic pattern is viewed through both eyepieces. While ranging it is not necessary to note the range indicated on the range scale, since it is automatically set into the range finder during this operation.

c. Provisions are made in the range finder to transmit data electrically to the Superelevation Transmitter, M22 (see appendix III). This instrument is mounted on the power pack and is connected by a cable to a receptacle on the back of the range finder. As the gunner ranges on the target, with the turret power controls on, the superelevation transmitter automatically causes the gun to be elevated or depressed to the required superelevation angle, and keeps the ranging pattern in contact with the target image during this operation. After ranging is completed, the gun is laid for firing by means of the turret traversing and gun elevating controls. Superelevation is the angle above the line of sight to which the gun tube must be elevated to hit a target at a given range.

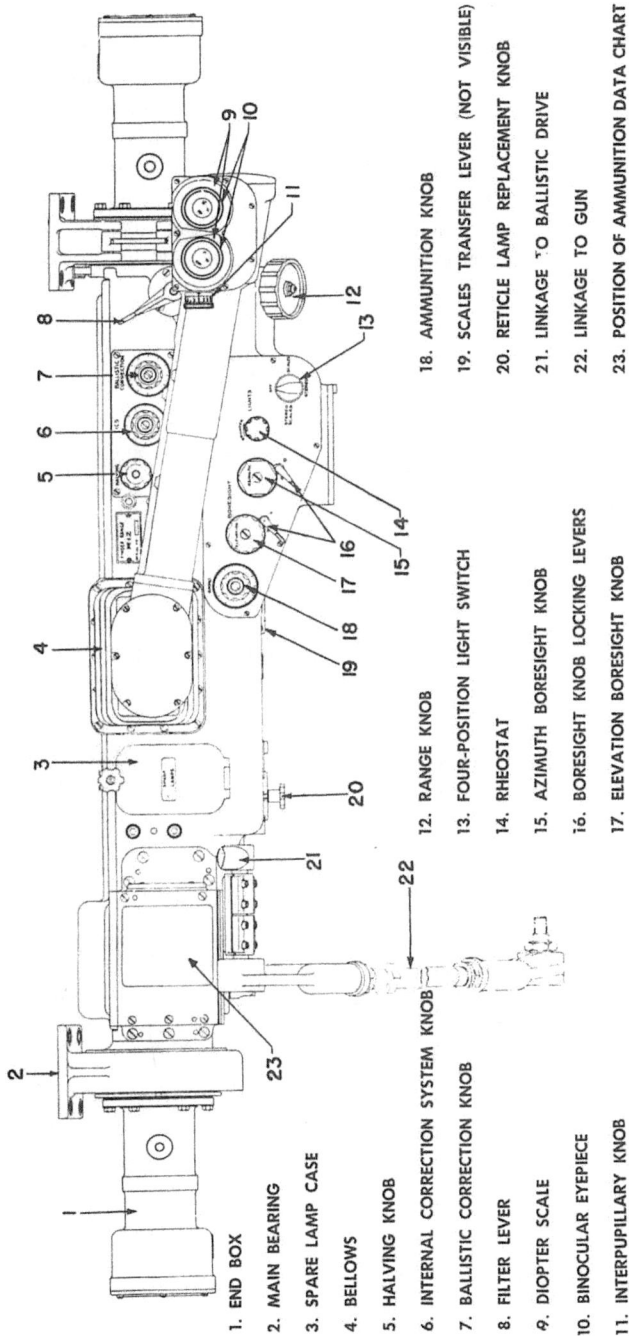

Figure 15. Range finder, M12.

1. END BOX
2. MAIN BEARING
3. SPARE LAMP CASE
4. BELLOWS
5. HALVING KNOB
6. INTERNAL CORRECTION SYSTEM KNOB
7. BALLISTIC CORRECTION KNOB
8. FILTER LEVER
9. DIOPTER SCALE
10. BINOCULAR EYEPIECE
11. INTERPUPILLARY KNOB

12. RANGE KNOB
13. FOUR-POSITION LIGHT SWITCH
14. RHEOSTAT
15. AZIMUTH BORESIGHT KNOB
16. BORESIGHT KNOB LOCKING LEVERS
17. ELEVATION BORESIGHT KNOB

18. AMMUNITION KNOB
19. SCALES TRANSFER LEVER (NOT VISIBLE)
20. RETICLE LAMP REPLACEMENT KNOB
21. LINKAGE TO BALLISTIC DRIVE
22. LINKAGE TO GUN
23. POSITION OF AMMUNITION DATA CHART

d. An Inverter, 7633698, is used in conjunction with the superelevation transmitter and range finder. It is located in the turret behind the radio and changes the 24-volt dc, supplied by the tank power source, to the 110-volt 400-cycle ac that is required to operate the synchros in the range finder and superelevation transmitter.

e. In the event an end box is shot out or damaged, means are provided to allow either side of the range finder to be used as an offset telescope. When the range finder is used in this manner, the target range is estimated and indexed on the range scale. If the left side of the instrument is damaged, the scales and gun-laying reticle in the left eyepiece may be transferred to the right eyepiece.

f. Data.

Operating temperature range_____ —65° F. to +150° F.
Operating voltage_____ 24 volts dc, and 110 volts, 400-cycle ac.
Elevation limit_____ +356 mils.
Depression limit_____ —178 mils.
Range _____ Graduated from 500 to 5,000 yards.
Magnification_____ 7.5x.
Field of view_____ 89 mils.
Range finder base length_____ 60 inches.

g. Reticle Patterns and Scales.

    (1) *Reticles and scales.* The following reticles and scales (fig. 16) will be visible to the observer when the light switch is indexed to STEREO SCALES and the scale transfer lever is

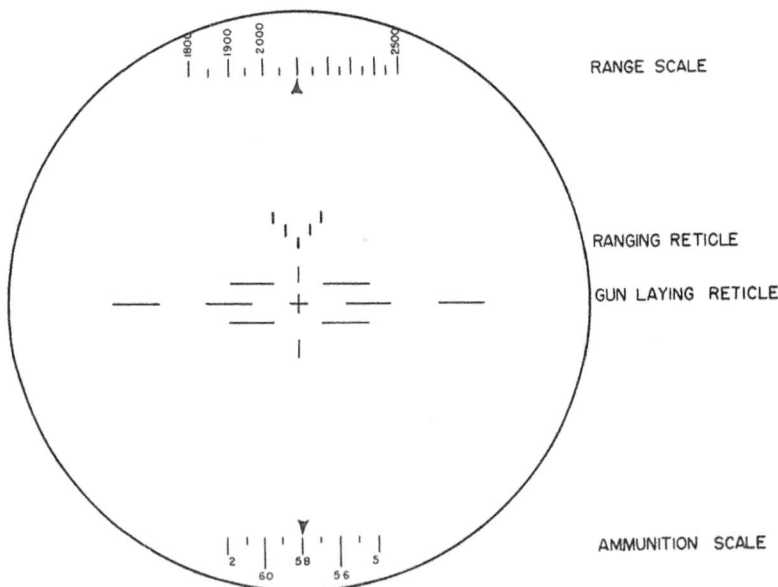

Figure 16. Reticle pattern and scales, range finder, M12.

positioned away from the observer. This is the normal carrying position.

(*a*) Range scale and index.

(*b*) Stereoscopic pattern (ranging reticle).

(*c*) Gun-laying reticle.

(*d*) Ammunition scale and index.

(2) *Range scale and index.* The range scale and index are seen in the top of the field of view. The range scale is graduated from 500 to 5,000 yards. The "B" at the left end of the scale must always be alined with the index when boresighting. The scale is moved past the index by rotating the range knob and indicates the range set in the range finder. If the range finder is being used as an offset telescope, the estimated range must be indexed on the range scale. The range scale is normally viewed through the left eyepiece.

(3) *Steroscopic pattern (ranging reticle).* One ranging reticle appears below the range scale in the left eyepiece. A similar ranging reticle appears in the right eyepiece. When the halving knob is properly adjusted, these reticles appear to be at the same elevation. Each reticle is formed by five vertical lines placed to form a ⌐|¬ pattern. This pattern is used to gain stereoscopic contact with the target. With the range finder in proper adjustment, the ranging reticles will fuse together and appear as one projected into space in front of the tank. In addition, the reticle appears to have depth: The lowest bar appears to be farther away from the gunner than the two center bars; the two upper bars appear to be closer than the two center bars. As the range knob is rotated to increase the range, this ranging reticle will appear to move away from the gunner. When ranging on a target, the range knob is rotated until the lowest vertical bar of the ranging reticle appears to be at the same range as the target. The range scale in the upper portion of the eyepiece will indicate the range to the target.

(4) *Gun-laying reticle.* The gun-laying reticle appears in the field of view below the ranging reticle. It is commonly called the "sagebrush" reticle (fig. 20). In the center of the reticle is the aiming cross. This cross is placed on the center of the target to fire the first round except when a lead is necessary.

(5) *Ammunition scale and index.* The ammunition scale and index appear in the bottom of the field of view. The ammunition scale is graduated from 0 to 90. An ammunition data chart is mounted on each range finder, indicating the proper ammunition code numbers for the various types of ammuni-

tion. When ammunition is announced by the tank commander in the initial fire command, the gunner will rotate the ammunition knob until the proper code number for the type of ammunition to be fired is indexed on the ammunition scale. The scale must always be indexed at 0 when boresighting.

h. Controls.

(1) *Diopter adjustment.* The diopter adjustment, which is provided for each eyepiece, is used to focus the eyepieces to the gunner's eyes. The diopter scale is located around each eyepiece and is graduated from −3 to +3 diopters. Adjustment or focusing is accomplished by rotating the eyepiece.

(2) *Interpupillary adjustment.* The interpupillary adjustment knob is provided so that the distance between the eyepieces can be adjusted to match the distance between the gunner's eyes. The adjustment knob is located to the left of the eyepiece and is graduated from 58 to 72 millimeters in 1-millimeter increments.

(3) *Halving knob.* The halving knob is on the left side of the upper control panel. This knob permits vertical adjustment of the left ranging reticle. The halving adjustment is made when the left and right ranging reticles appear at different elevations within the field of view.

(4) *Internal correction system knob.* The internal correction system knob (ICS) is in the center of the upper control panel. Around the knob is a scale graduated from 0 to 50 in units of error. This control permits the adjustment of the instrument's internal mechanism to the eyes of the gunner. The setting will vary for each instrument and gunner. See paragraph 33c(10) for procedure for determining ICS setting.

(5) *Ballistic correction knob.* The ballistic correction knob on the right side of the upper control panel is designed to correct for variables that affect the muzzle velocity or terminal velocity of the various types of ammunition fired in the 90-mm Gun, M36. The knob is graduated from −11 to +16 in percentage of the range. Muzzle velocity or terminal velocity may be changed by tube wear, air temperature, wind, different ammunition lots, and other factors. In the absence of ballistic correction data, the ballastic correction knob will be set at zero.

(6) *Filter lever.* The filter lever is located to the right of the upper control panel. It controls filters for the left and right optical systems of the range finder. When the lever is to the right, the filters are removed from the gunner's field of view; when the lever is moved to the left, the filters are introduced into the optical system.

(7) *Ammunition knob.* The ammunition knob, on the left side of the lower control panel, controls the movement of the ammunition scale seen in the lower portion of the field of view. The ammunition scale is moved by rotating the ammunition knob until the desired code number is indexed. Ammunition numbers are determined by referring to the ammo data chart on the left side of the range finder.

(8) *Elevation boresight knob.* The elevation boresight knob is located to the right of the ammunition knob on the lower control panel. It is used to control vertical movement of the gun-laying reticle during boresighting and zeroing. The adjustable slip scale around the elevation boresight knob is graduated from 0.5 to 5.5 mils at 0.1-mil intervals, and can be rotated independently of the knob. A locking lever on the elevation boresight knob insures that the desired setting will not be changed by vibration during operation of the tank.

(9) *Azimuth boresight knob.* The azimuth boresight knob is located to the right of the elevation boresight knob on the lower control panel. It is used to control the horizontal movement of the gun-laying reticle during boresighting and zeroing. The adjustable slip scale around the azimuth boresight knob is graduated from 0.5 to 5.5 mils and can be rotated independently of the knob. The locking lever performs the same function as that on the elevation boresight knob.

(10) *Light switch.* The light switch is a four-position switch located on the right side of the lower control panel. Four positions around the switch are marked clockwise as follows: OFF, SCALES, STEREO, and STEREO SCALES (carry position). The position of the switch controls the illumination of various scales and reticles as follows:

	Off (12 o'clock)	Scales (3 o'clock)	Stereo (6 o'clock)	Stereo scales (9 o'clock)
Range scale_____		X		X
Ranging reticle_____			X	X
Gun laying reticle _____		X		X
Ammo scale_____		X		X

Brightness of the reticles and scales is adjusted by rotating the rheostat knob located to the left of the light switch.

(11) *Range knob.* The range knob is located below and in front of the eyepieces. It is used to control the movement of the range scale and the ranging reticle. By rotating the range

knob, the range scale can be moved through all graduations from 500 yards to 5000 yards. This rotation also will cause the ranging reticle to appear to move away from or toward the gunner as he looks through the eyepieces.

(12) *Scale transfer lever.* The scale transfer lever is located in the center and forward on the bottom of the range finder. It is used to transfer the gun-laying reticle, range scale, and ammunition scale from the left to the right optical system. This is done when the left end box is damaged and the range finder is being used as an offset telescope. The normal position for the lever is away from the gunner so that the scales and reticles appear in the left optical system. To transfer the scales, pull the lever to the rear and toward the gunner. This lever is spring loaded, and it must be eased into position to prevent damage to the range finder. When the scale transfer lever is positioned toward the gunner, the ranging reticles are not visible.

(13) *Right reticle lamp.* The right ranging reticle lamp is located to the left of the scale transfer lever. If this lamp burns out, it can be replaced by loosening the knurled knob and allowing the right reticle lamp door to swing open; this makes the lamp accessible for replacement.

(14) *Scale and left reticle lamps.* The scale and left ranging reticle lamps are located to the right of the scale transfer lever. These lamps may be replaced in the same manner as the right ranging reticle lamp.

(15) *Spare lamps.* Spare lamps for the left ranging reticle, right ranging reticle, and scales are located in the spare lamp box on the rear of the range finder. Each range finder is initially equipped with eight spare lamps.

> *Note.* In replacing scale and reticle lamps, insure that the silvered half of the lamp is down and the frosted portion up when the lamp housing door is closed.

i. Operation. The following steps are necessary to put the range finder into operation. Due to the functional relationship of the various controls, it is recommended that these steps be performed as listed.

(1) *Diopter adjustment.* Turn both eyepieces to index maximum plus settings. Cover one eyepiece. With both eyes open, slowly rotate the uncovered eyepiece until the image is sharp. Repeat for the other eyepiece. Do not rotate past the point where the image first appears to be sharp.

(2) *Interpupillary adjustment.* The gunner applies his known interpupillary distance on the interpupillary scale. If this setting is not known, use the interpupillary distance deter-

mined with the binocular (see paragraph 95a, FM 17-12). This will provide an approximate setting for the range finder.

Note. Improper interpupillary adjustment will prevent the gunner from seeing the ranging reticle in depth.

(3) *Scale transfer lever.* The gunner checks the scale transfer lever to insure that it is positioned away from him.

(4) *ICS adjustment.* The gunner places his individual ICS setting on the ICS scale. This introduces the gunner's personal correction into the range finder. In the event the gunner has not determined his own ICS setting (par. 33c(10)), he uses a setting of 25.

(5) *Lights and filter adjustment.* The gunner turns the four-position light switch to the STEREO-SCALES position. He introduces the filter if desired. Next, he adjusts the brightness of the reticles and scales with the rheostat. The intensity of light should be reduced to the minimum which will still permit the use of the reticles and scales.

(6) *Halving adjustment.* Each time the gunner places the range finder into operation, he will make his halving adjustment. It can be done by either of the following methods. In the first, or binocular, method the gunner sets a range of 5000 yards on the range scale. He depresses the range finder to look at the ground just ahead of the tank; as this is done, the ranging reticle will break up into a left and right reticle. Then, with both eyes open, he moves the left ranging reticle with the halving knob until it appears to be at the same elevation as the right ranging reticle. In the second method, he uses the gun controls to lay the right ranging reticle at a given elevation on a target at about 1500 yards. He closes his right eye, opens his left, and lays the left ranging reticle at the same elevation on the target, using the halving knob. He then closes and opens his eyes alternately to recheck adjustment. The first method of making the halving adjustment is the more accurate.

(7) *Sight adjustment.* The gunner will perform the steps of boresighting and zeroing (pars. 31 and 32).

(8) *Battle sights.* A battle sight is a predetermined range and ammunition setting applied to the sights for use when firing at dangerous surprise or fleeing targets. A battle sight is designed for use in an emergency and should be specified in the unit SOP. It is determined on the basis of experience, intelligence, and/or an analysis of the terrain and enemy situation. A typical battle sight might be armor-defeating (SHOT) ammunition and 800 yards range.

(*a*) The ammunition setting will correspond with the type of ammunition in the gun or expected to be needed for the next target.

(*b*) The range scale will be set to indicate the normal range of employment for the terrain and mission.

(*c*) Both settings are variable and will be determined by the tank unit for its particular situation.

j. Procedure for Installing End Boxes.

(1) If an end box is shot out or damaged, it can be replaced by the turret artillery mechanic. The procedure for performing this task is described below.

(2) Before mounting the end box on the main housing, check both mounting surfaces to see that they are free from dirt or burrs. Clean both surfaces thoroughly.

(3) Carefully place the end box against the main housing, applying pressure in the direction which will force a stationary key in the end housing against the key slot in the main housing.

(4) Insert the six bolts to hold the end box to the main housing and tighten with *finger presssure only*. Check the end box to make certain that it does not rock on the main housing in any direction.

(5) Make certain that the wedges are loose in the key slots of the main housing, and then tighten the wedges with a torque wrench. Alternate from top to bottom, using 10 inch-pounds on the top, 20 inch-pounds on the bottom, 30 inch-pounds on the top, and 40 inch-pounds on the bottom; finish with 40 inch-pounds on the top and bottom.

(6) Tighten the retaining bolts on the end housing in the following manner (fig. 17):

Left end box	Right end box
2—30 inch-pounds	1—90 inch-pounds
5—30 inch-pounds	5—90 inch-pounds
3—60 inch-pounds	2—90 inch-pounds
6—60 inch-pounds	6—90 inch-pounds
4—90 inch-pounds	3—90 inch-pounds

30. Periscope, M20, and Ballistic Drive, M3

a. The secondary direct-fire control equipment is composed of the Ballistic Drive, M3, with Instrument Light, M30; Gunner's Periscope Mount, M88, and Periscope, M20, with Instrument Light, M36; Commander's Periscope Mount, M89, and Periscope, M20, with Instrument Light, M36 (fig. 18).

b. The Periscope, M20 (see appendix III), a single-eyepiece type instrument, has two built-in optical systems: a one-power system (ob-

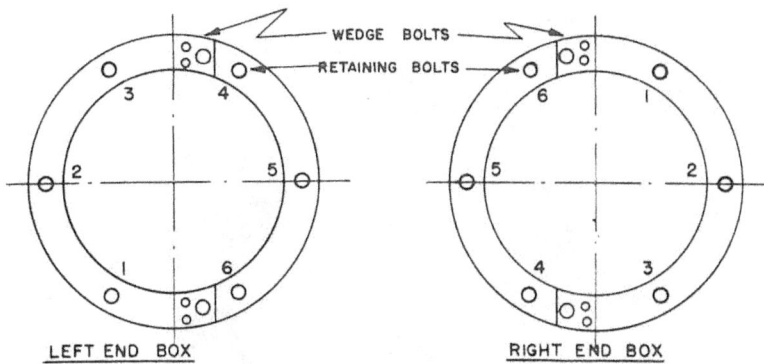

Figure 17. Installation of end boxes.

servation window) for wide-angle, close-in observation, and a six-power system for sighting. The periscope is composed of two separate parts: a head assembly which is supported at the top of the periscope mount, and a body assembly (fig. 19) which is secured to the bottom of the mount.

(1) The Commander's Periscope, M20, serves as a target-designating device. A coupling assembly connects the periscope head to an extension arm on the ballistic drive. The periscope head contains a movable mirror that follows the movement of the gun as it is elevated or depressed. The six-power system of the periscope contains a sight reticle. In an emergency, the tank commander may use his M20 periscope as a direct fire sight.

(2) The Gunner's Periscope, M20, is the secondary direct fire sight. It is supported on the Gunner's Periscope Mount, M88. A coupling assembly with an elevation lever connects the movable mirror in the periscope head to a coupling on the shaft of the ballistic drive. The shaft of the ballistic drive is also connected to the gun linkage by means of an offset connector. This connector is used to link the ballistic drive to the gun trunnion only when the range finder is removed from the tank. When the range finder is installed, the offset connector is removed and stored with second echelon supplies. The ballistic drive is then linked to the gun through the range finder (fig. 18). When the gun is elevated or depressed, the mirror in the periscope head moves, causing the gunner's line of sight through the periscope to be elevated or depressed the same amount as the gun. The line of sight of the periscope is also moved independent of the gun when the knob on the ballistic unit of the ballistic drive is rotated.

41

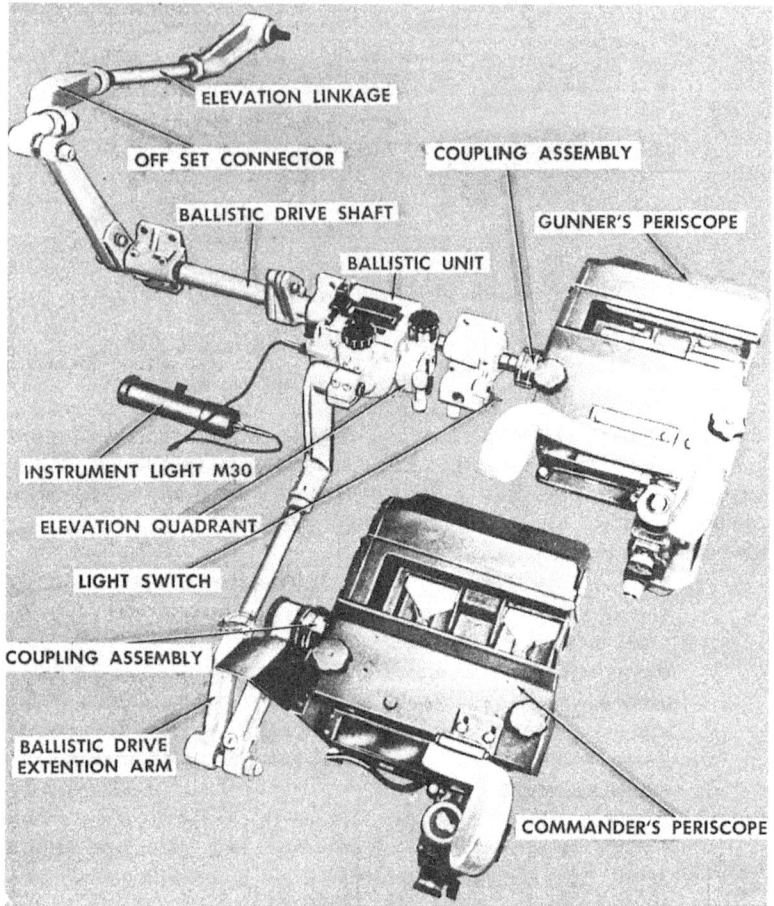

Figure 18. Arrangement of ballistic drive, M3; gunner's periscope mount, M88; and commander's periscope mount, M89, without range finder installation.

(3) The reticle pattern ("sagebrush" reticle, fig. 20) in the Periscope, M20, six-power system is 40 mils wide, with each horizontal line measuring 5 mils. Each verticle line measures 2 mils. The aiming cross in the center of the reticle, with intersecting lines measuring 2 mils, is used for boresighting and for firing the initial round at stationary targets. The one-power system (observation window) of the Periscope, M20, has no reticle pattern.

(4) The lever on the coupling allows the line of sight through the periscope to be elevated independently as much as 22° from zero elevation. This increases the amount of terrain visible

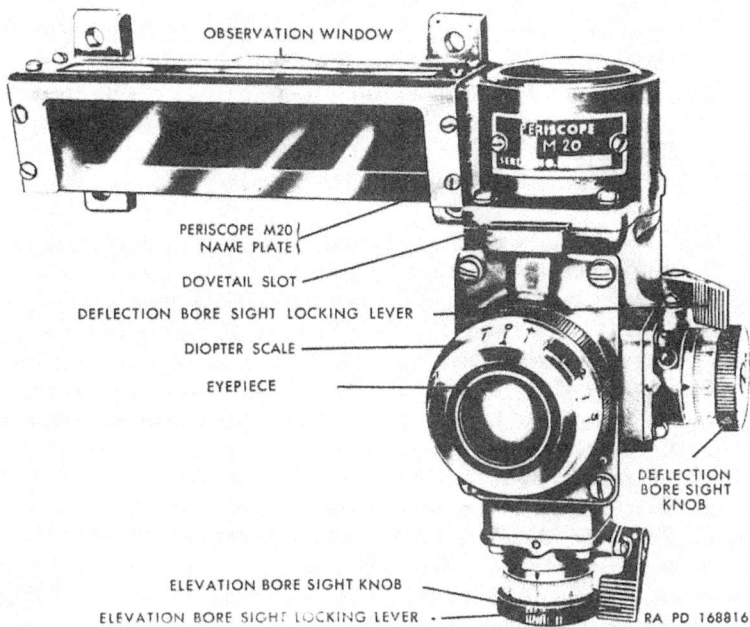

Figure 19. Body assembly, periscope, M20.

OBSERVATION WINDOW

PERISCOPE M20
NAME PLATE

DOVETAIL SLOT

DEFLECTION BORE SIGHT LOCKING LEVER

DIOPTER SCALE

EYEPIECE

DEFLECTION
BORE SIGHT
KNOB

ELEVATION BORE SIGHT KNOB

ELEVATION BORE SIGHT LOCKING LEVER

RA PD 168816

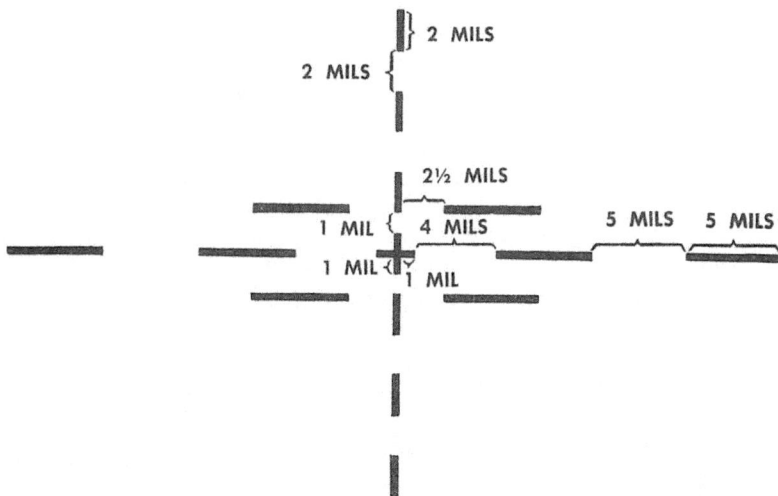

2 MILS

2 MILS

2½ MILS

1 MIL

4 MILS

5 MILS

5 MILS

1 MIL

1 MIL

Figure 20. Reticle pattern, periscope, M20.

43

through the periscope without elevating the gun. The lever must be held in the depressed position when installing the periscope head.

(5) A diopter adjustment is provided on the eyepiece of the six-power system for focusing the periscope to the eye of the user. The diopter scale is graduated from -3 to $+3$ diopters. This facilitates resetting the adjustment once the proper setting is known.

(6) Two boresight knobs are provided on the periscope body for making reticle adjustments. Each knob has a locking lever to hold the desired adjustment, and a slip scale graduated from .5 to 5.5 mils for recording the adjustments made in boresighting and zeroing.

(7) Reticles of the Periscope, M20, are illuminated by means of the Instrument Light, M36. A dovetail slot over the eyepiece of the six-power system receives the lamp bracket of the instrument light.

(8) The Gunner's Periscope Mount, M88, and Commander's Periscope Mount, M89 (see appendix III), are used to support the Periscopes, M20. The commander's periscope mount is attached to the front of the commander's cupola. The center front direct vision block in the commander's cupola has been removed and the commander's M20 periscope mounted in this opening. A splash guard is installed on the back of each mount for protection against high-explosive fragments and small-arms projectiles in the event the periscope head is hit. Two retaining knobs hold the guard on the mount. The gunner's mount has a full headrest, and the commander's mount has a half headrest. A cable from the loader's reset safety is connected to the receptacle at the base of the periscope mounts. This supplies power to light the gun's ready light, and an illuminating tube reflects the light to the gunner's and commander's viewing positions.

c. The Ballistic Drive, M3 (see appendix III), is composed of a ballistic unit and an arm assembly. The ballistic unit introduces superelevation, for the range and type of ammunition being used, when the periscopes are used for laying the weapon. The arm assembly connects the Periscope, M20, to the ballistic unit and also to the gun linkage through the range finder.

d. The ballistic unit contains a range drum with range scales graduated for different types of ammunition. The ammunition scales in the ballistic unit provide for a selection of the various types of ammunition. There is also a mil scale on which, with the use of a firing table, the correct elevations for ammunitions not included on the ammunition scale may be set into the ballistic drive. The range scale

44

is controlled by rotating the range knob. A light switch on the left side of the ballistic unit is the control for the illumination of the range scales and ammunition scale.

31. Boresighting

a. *General.*

(1) Position the tank as level as possible.

(2) Insert the breech boresight into the chamber, or remove the percussion mechanism and use the firing pin well as a breech boresight. Tape black thread to form a cross over the witness lines on the muzzle end of the tube.

(3) Perform the following steps to place the range finder into operation: Adjust diopter setting; make interpupillary adjustment; check position of scale transfer lever; place individual ICS setting on the ICS knob; check scale and reticle lights; position filter, if needed; and adjust halving. (See paragraph 29*i*.)

(4) Select a target as near 1500 yards as possible (preferably a target with clearly defined horizontal and vertical lines).

(5) Position the right telescope of a binocular over the firing pin well. Using the firing pin well (breech boresight) as a rear sight and the cross threads on the muzzle as a front sight, aline the axis of the bore on the boresight point by use of the manual traverse and elevation controls.

> *Note.* Make sure that the accumulator pressure is high enough to hold the gun on the boresight point. (See paragraph 21*c*.)

b. *Range Finder, M12.*

(1) Rotate the ballistic correction knob to 0.

(2) Rotate the ammunition knob until 0 is indexed on the ammunition scale or knob.

(3) Rotate the range knob until the letter "B" is indexed on the range scale.

(4) Unlock the elevation and azimuth boresight knobs. While looking through the eyepieces, manipulate the boresight knobs to place the aiming cross of the gun-laying reticle on the precise point upon which the bore of the gun has been alined. Relock the boresight knobs.

(5) Recheck the bore alinement to insure that the tube remains alined on the boresight point.

(6) With the boresight knobs locked, move the slip scale on the elevation boresight knob to index "4" and the slip scale on the azimuth boresight knob to index "3." This completes the steps of boresighting the range finder.

c. *Periscope, M20.*

    (1) Set the range scale in the ballistic unit at 0 by turning the range knob counterclockwise until its stops.

    (2) Sighting through the commander's and gunner's periscope eyepieces, aline the aiming cross of each reticle exactly on the boresight points. Make this alinement by rotating the elevation and deflection boresight knobs without disturbing the lay of the gun.

    (3) Clamp the boresight knobs with the locking levers, and turn the slip scale to read "4" on the elevation boresight knob and "3" on the deflection boresight knob.

32. Zeroing

    a. *Boresighting.* Boresight as in paragraph 31.

    b. *Range Finder, M12.*

    (1) Unlock the elevation and azimuth boresight knobs. Then, being careful not to move the slip scales, rotate each knob until the figure "2" is indexed. Relock the boresight knobs.

        Note. This is the emergency zero of the range finder. Whenever possible, the range finder should be zeroed as described below.

    (2) Rotate the ammunition knob to index the code number for the type of ammunition which is to be used for zeroing. (Shot ammunition should be used.)

    (3) Rotate the range knob until the exact range to the target is indexed on the range scale.

    (4) Fire a shot group of from three to five rounds on the target. Check the lay of the gun after each round, and re-lay if necessary.

        Note. The first round should be on the target. If it is not, use the "burst-on-target" principle of adjustment to hit the target (par. 95). Using the elevation and azimuth boresight knobs, move the aiming cross to this hit. Aiming at this point, fire the shot group.

    (5) Without disturbing the lay of the gun, unlock the elevation and azimuth boresight knobs. Move the aiming cross of the gun-laying reticle to the center of the shot group by manipulating the boresight knobs. Relock the boresight knobs.

    (6) Fire a round to check the zero. If the check round strikes within 14 inches of the point of aim, the gun is zeroed.

    (7) Record the final reading on the elevation boresight knob and azimuth boresight knob.

        Note. The final readings on the boresight knobs become the established zero for that particular gun and range finder, and should be recorded at some convenient place in the tank turret. They may also be recorded in pencil in the artillery gun book. This zero should be verified periodically by firing a check round.

c. Periscope, M20.

(1) Unlock the elevation and deflection boresight knobs. Rotate each knob until the figure "2" is indexed. Relock the boresight knobs. This is the emergency zero for the Periscope, M20.

(2) Turn the range knob until the range scale indicates the known range to the target for the type of ammunition to be fired.

(3) Using the elevation and deflection boresight knobs, lay the aiming cross on the center of impact established above. Lock the boresight knobs.

(4) When the check round is fired, it should strike within 14 inches of the point on the target where the aiming crosses of the reticles of the gunner's and commander's periscopes are laid. If so, the Periscopes, M20, are zeroed, and the settings on the boresight knobs are the zero settings for the periscopes.

33. Training of Stereoscopic Range Finder Gunners

a. General. The range finder is a precision optical instrument. In order to get the ultimate from this instrument, all phases of training must emphasize this point. In addition to being a range-determining instrument, the range finder is the primary direct-fire sight on the M47 tank. The gunner must be capable of rapid as well as accurate fire with this instrument. The time required for a trained gunner to determine the range to any target accurately must not exceed 5 seconds and should be even less. In determining ranges with the M12 range finder, accuracy cannot be sacrificed for speed. Instead, both must be developed simultaneously.

b. Equipment and Areas Necessary for Training.

(1) The following throw-over charts, or similar visual aids, are desirable for unit training on the M12 range finder:

(*a*) M20 periscope and M12 range finder, complete with linkage.

(*b*) Rear view of M12 range finder.

(*c*) Blow-up of M12 range finder control panel.

(*d*) Three views of range finder reticle showing both eyepieces with:

1. Lights on STEREO.

2. Lights on SCALES.

3. Lights on STEREO-SCALES.

(*e*) Three stereo composites showing ranging reticle suspended over target with range set:

1. Short.

2. Over.

3. At target range.

(2) The training area for range finder instruction must include a target-ranging area with ranges up to at least 3,500 yards.

Targets should be placed in this area at known ranges from 700 yards to the maximum range available. There should be between 12 and 24 targets in all available types of locations, such as on forward slopes (steep and gradual), on the skyline, partially visible over a ridge line, against contrasting background, in thick vegetation, and in sparse vegetation. Some of the targets should be partially camouflaged. For preliminary training, targets with sharp outlines and little or no depth are desirable. As skill in ranging is gained, the gunner can range on the more difficult targets accurately.

c. *Training Procedures and Techniques.*

(1) Physical examination of personnel prior to inauguration of range finder training is not necessary, but it is an aid in eliminating those who are physically incapable of operating a stereoscopic range finder. This is of particular importance in units that are short of equipment. The examination would consist of a test using an orthorator or similar device which will determine whether the trainee can see stereoscopically. The examination should also include a measurement of the interpupillary distance of each man. The interpupillary adjustment on the M12 range finder will accommodate measurements from 58 to 72 millimeters.

(2) If physical examination of personnel by a medical officer is not possible, men falling into the following categories can still be eliminated prior to inauguration of training:

(*a*) Those who have vision in only one eye.

(*b*) Those who suffer serious muscular impairment of one eye.

(*c*) Those whose interpupillary distance is less than 58 millimeters or greater than 72 millimeters.

(*d*) Those who demonstrate physical, mental, or moral deficiencies of a nature to render them incapable of performing the duties of a gunner or tank commander.

(3) Any conference on the range finder should cover use, operation, maintenance, and adjustment. The adjustment and/or setting of the following will be discussed: diopter scales, interpupillary scale, halving knob, boresight adjustment knobs with slip scales, ICS knob, ballistic correction knob, light switch and rheostat, range knob, filter lever, scale transfer lever, reticle lamps, and spare lamps.

(4) When gunners start working with the range finder, they must be checked individually to make sure they can see the ranging reticle stereoscopically.

(5) When the gunner can see the ranging reticle stereoscopically, practice will develop the required speed and accuracy of ranging. During early stages of training, long periods of

ranging should be avoided. It is desirable to conduct ranging practice in 4-hour periods. This ranging practice must be very closely supervised. The practice rangings must be accomplished with the light switch on STEREO. When the gunner has completed his ranging on a target, he moves his head aside and turns the light switch to SCALES. A man sitting in the tank commander's seat can now lean forward, read the range set on the range finder, and call it off to a recorder. The gunner now switches back to STEREO, traverses to the next target, and ranges. The targets should be engaged in the order of 1, 2, 3, 1, 2, 3, 1. The targets should be located in a limited sector of the range area to avoid interference of gun tubes as adjacent tanks range on different targets. As soon as the gunner is familiar with the range finder, he must work for speed by trying to obtain as many readings as possible during his period of ranging.

(6) Initially, the gunner should be allowed 30 seconds to complete one ranging. Ranging time should be reduced by 5 seconds during each ranging period thereafter until his ranging time on each target is a maximum of 5 seconds.

(7) In ranging on a target, the gunner must be taught to move the ranging reticle toward the target until the target appears to be bracketed in range between the lowest vertical bar of the reticle and the two center bars. Then he makes his final adjustment, moving the reticle toward himself until the lowest bar appears to be at the same range as the target.

(8) A new block of targets should be assigned each hour. The control radio will be used to control ranging periods, to rotate gunners, to designate blocks of targets to be used, and to notify personnel of breaks. Although some personnel will qualify after a few hundred rangings, it normally takes 2000 to 3000 readings for a gunner to reach his maximum efficiency.

(9) Records must be kept of all practice readings. These are necessary to tabulate the number of rangings obtained by each gunner and to determine his progress in obtaining accuracy. The more rangings a man makes, the more rapid and more accurate he becomes. The learning period will vary with individuals.

(10) As a man becomes proficient, his ICS setting can be determined and applied to the instrument so his average range to a target will be the true target range. This ICS setting will vary by instrument and must be determined each time a gunner uses a new instrument. To determine his ICS setting, the gunner selects a target of known range as near 1500 yards

as possible. He sets this known range to the target on the range scale. He then turns the ICS knob, causing the ranging reticle to move toward him or away from him until the lower bar of the ranging reticle appears to be at the exact same range as the target. He notes and records the reading that is set on the ICS scale. He repeats this procedure until he has at least ten settings recorded. He finds the average of these settings and uses that figure as his personal ICS setting. Since the ICS setting may vary slightly from day to day, the gunner should check his ICS setting frequently.

(11) The training of stereoscopic range finder gunners should be pointed toward reducing the gunner's deviation in ranging on a 1500-yard target to four units of error (UOE). Expressed in yards, the operator's average error in ten rangings of a 1500-yard target should be 42 yards or less. This is derived by applying the unit-of-error formula as follows:

$$4.65 \times \left(\frac{\text{Range}}{1000}\right)^2 = 1 \text{ UOE in yards}$$

EXAMPLE

1500 yards

$$4.65 \times \left(\frac{1500}{1000}\right)^2 = 4.65 \times (1.5)^2$$

or

$$4.65 \times 2.25 = 10.46 \text{ yards or } 10.5 \text{ yards}$$

Four units of error at 1500 yards equals $4 \times 10.5 = 42$ yards.

34. Synchronization and Backlash

a. Synchronization. Using units are not authorized to adjust the linkage arms between the gun trunnion, periscopes, and range finder. They may check the linkage adjustment as follows:

(1) Boresight on a target at least 5 miles distant.
(2) Place the tank on a steep forward slope and check the boresight when the gun is near maximum elevation.
(3) Place the tank on a steep reverse slope and check the boresight when the gun is near its maximum depression.
(4) If the range finder or periscope is not accurate within .5 mils with the gun at maximum elevation or depression, the linkage should be adjusted. Linkage adjustment is an ordnance function.

Note. The actual adjustment of the linkage must be performed by ordnance personnel.

b. Backlash. A backlash check with all fire-control instruments securely locked in place may be performed as follows:

50

(1) Starting with the gun elevated above a target at a range of as near 1500 yards as possible, depress the gun until the aiming cross is on the aiming point. Carefully measure the existing elevation with the M1 gunner's quadrant.

(2) Depress the gun below the aiming point, and then elevate it until the aiming cross is again alined precisely on the aiming point. Measure the existing elevation with the gunner's quadrant.

(3) The difference between the two quadrant readings is the backlash in the system. It should not exceed 0.3 mil.

Note. The effects of backlash can be largely eliminated if the gunner makes the final lay of the gun from the same direction each time.

35. Vision Devices

The driver, bow gunner, and loader in the M47 tank have plain window or plastic Periscopes, M6 or M13, for observation. In addition, the tank commander's cupola is provided with vision blocks to permit all-around vision with the hatch closed.

36. Auxiliary Fire-Control Equipment

a. General. The Tank, 90-mm Gun, M47, is equipped with fire-control equipment for indirect laying of the tank gun when the target is not visible to the gunner. This equipment includes the Azimuth Indicator, M31, and the Elevation Quadrant, M13.

b. Elevation Quadrant, M13.

(1) The Elevation Quadrant, M13 (fig. 21) (see appendix III), is mounted on the right side of the ballistic unit. It is used to measure vertical angles and to provide a means for setting mil elevation on the gun.

(2) It is composed of a micrometer knob, micrometer scale and index, elevation scale and index, reflector, level vial, and cover.

(3) The elevation scale is graduated from −200 to +600 mils. It is marked every 100 mils and numbered every 200 mils from 0 to −2 and from 0 to +6.

(4) The micrometer scale is graduated in both directions from 0 to 100 mils in 1-mil increments and is numbered every 10 mils from 0 to 90. The black figures are used for plus elevations and the red figures are used for minus elevations.

(5) One complete revolution of the micrometer knob moves the elevation scale index 100 mils on the elevation scale.

(6) The level vial and reflector move with the elevation scale index when angles are set on the instrument.

(7) The elevation quadrant moves with the gun. When the bubble is centered in the level vial, the gun is laid at the angle of elevation or depression set on the instrument, provided the ballistic unit is set at zero range.

51

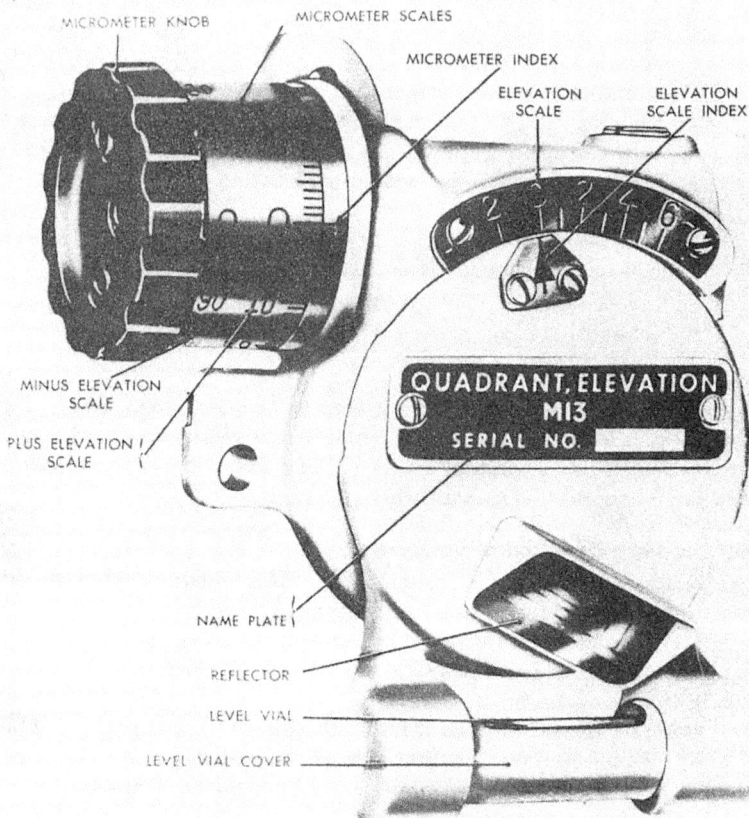

MICROMETER KNOB

MICROMETER SCALES

MICROMETER INDEX

ELEVATION
SCALE

ELEVATION
SCALE INDEX

MINUS ELEVATION
SCALE

PLUS ELEVATION
SCALE

QUADRANT, ELEVATION
M13
SERIAL NO.

NAME PLATE

REFLECTOR

LEVEL VIAL

LEVEL VIAL COVER

Figure 21.   Elevation Quadrant, M13.

(8) The reflector over the level vial provides a means for the gunner to more easily see the bubble in the level vial.

(9) Illumination of the level vial, elevation scale, and micrometer scale is provided by the Instrument Light, T22. The light switch, located to the right of the M13 elevation quadrant, has three positions: center—off; up—3-volt power source; down—24-volt power source. The up position is used for emergency operation when the tank's 24-volt system is not in operation.

(10) The adjustment of the M13 elevation quadrant can be checked by means of the M1 gunner's quadrant. Level the gun, using the M1 gunner's quadrant. Set the ballistic unit at zero range. Without disturbing the lay of the gun, center the bubble in the level vial of the M13 quadrant by rotating the micrometer knob. Check the elevation scale. If 0 is not

indexed on this scale, loosen the screw at each end of the scale and slip it until 0 is opposite the elevation scale index. Tighten the screws. Check the micrometer scale. If 0 is not indexed, loosen the three screws on the micrometer knob, then slip the micrometer scale. Check the bubble to be sure it is still centered in the level vial; if it is, tighten the three screws on the micrometer knob, and the instrument is ready for use. If the bubble is not centered, repeat the preceding instructions.

Note. When the Elevation Quadrant, M13, is used to measure or set off vertical angles, the ballistic unit must be set at zero range.

c. Azimuth Indicator, M31.

(1) The Azimuth Indicator, M31 (fig. 22) (see appendix III), which is mounted on the right side of the turret with gears in mesh with the turret ring, measures horizontal angles of traverse. It is used principally for laying the gun for indirect fire. The azimuth indicator is a dialed instrument with pointers indicating the readings on the scales.

(2) The Azimuth Indicator, M31, has three scales and three pointers. The azimuth scale is graduated in 100-mil intervals and is numbered every 200 mils from 0 to 3200 counterclockwise in two consecutive semicircles around the scale. The micrometer scale is graduated counterclockwise in 1-mil intervals and numbered every 5 mils from 0 to 100. The gunner's aid is graduated in 1-mil intervals and numbered every 5 mils from 0 to 50 mils right and left. The directional pointer is fixed in relation to the longitudinal axis of the tank and gives a course reading on the azimuth scale. This reading indicates the number of mils the gun has traversed from the longitudinal axis of the hull. The azimuth pointer works in conjunction with the micrometer pointer. These pointers are adjustable and may be set at zero (by depressing and turning the resetter knob) when the sights are set on any desired reference point. The sum of these two pointer readings then gives a precise reading of the number of mils the gun traverses from the reference point. The azimuth and micrometer scales are fixed, while the gunner's aid dial may be rotated to any position. The gunner's aid is used in making deflection corrections by rotating the dial until the zero coincides with the position of the micrometer pointer. Right and left deflection corrections are then laid off on the gunner's aid dial. After the correction has been applied, move the zero of the gunner's aid to the micrometer pointer.

53

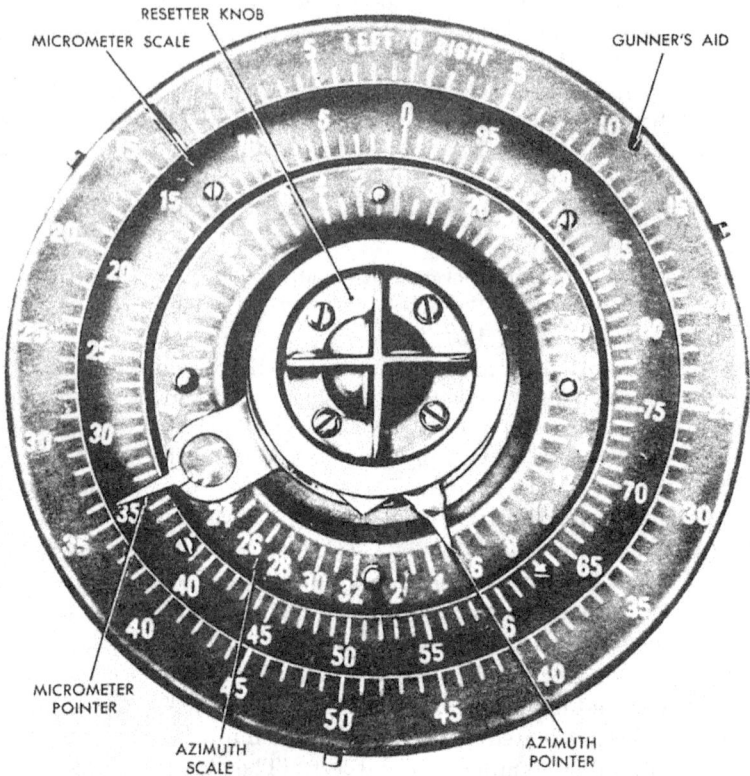

RESETTER KNOB

MICROMETER SCALE

GUNNER'S AID

MICROMETER
POINTER

AZIMUTH
SCALE

AZIMUTH
POINTER

Figure 22. Azimuth Indicator, M31.

(3) Built-in electric lamps provide illumination for the scales of the azimuth indicator. A receptacle on the side of the azimuth indicator receives the plug on the instrument light, which is installed in a bracket immediately above the indicator. The lamps, which are in the indicator, are turned on and off by a toggle switch on the instrument light.

(4) To test the accuracy of the azimuth indicator, lay the aiming cross of the periscope or range finder on a definite aiming point, and set the azimuth and micrometer pointers at zero. Traverse the turret manually through a complete circle until the sight is laid back on the original aiming point. If the azimuth and micrometer pointers do not indicate zero, the azimuth indicator is out of adjustment and needs a check by ordnance personnel.

(5) To test the azimuth indicator for slippage, lay the aiming cross of the sight on a definite aiming point, and set the azi-

muth and micrometer pointers at zero. Traverse the turret rapidly in power, and stop suddenly; repeat this operation two or more times in the same direction. Manually traverse the turret back to the aiming point. If the azimuth and micrometer pointers do not indicate zero, the azimuth indicator is slipping and will require adjustment by qualified ordnance personnel. If the pointers indicate zero, repeat the procedure in the opposite direction.

(6) To maintain the azimuth indicator, keep it clean and covered when not in use. Any lubrication or adjustment must be done by ordnance personnel.

37. Gunner's Quadrant, M1

a. *Uses.* The Gunner's Quadrant, M1 (fig. 23), standard equipment for each tank, is used to check the Elevation Quadrant, M13; to lay the gun for elevation when firing at point targets where greater accuracy is needed than can be obtained with the Elevation Quadrant, M13; and to measure elevation, depression, and cant of the gun.

b. *Description.*

(1) The Gunner's Quadrant, M1, consists of a sector-shaped frame to which is pivoted an index arm and level-vial holder. A scale graduated from 0 to 800 mils, at 10-mil intervals, is on one side of the frame, and a scale graduated from 800 to 1600 mils is on the opposite side.

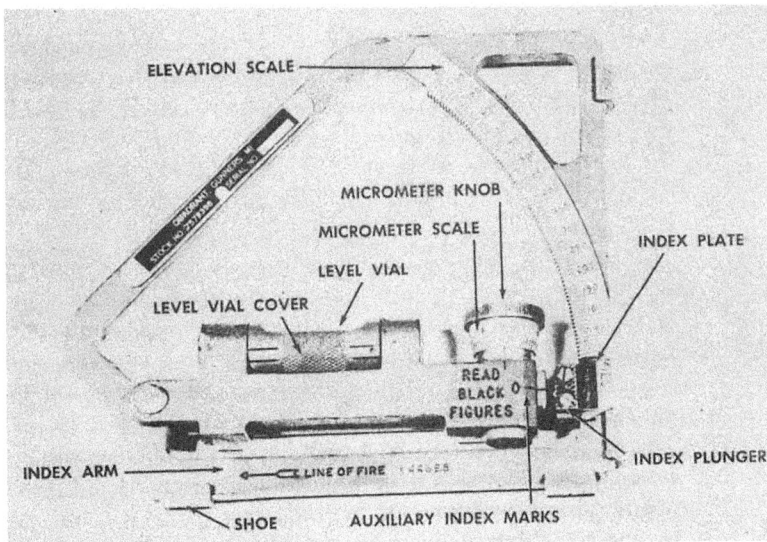

Figure 23. Gunner's quadrant, M1.

(2) The inside sector of the frame has teeth at 10-mil intervals. The teeth engage a spring-loaded, toothed plunger in the index arm and thereby permit setting of the arm to the desired angle as indicated on the scale.

(3) A level and micrometer mechanism are mounted on the index arm. The micrometer is graduated from 0 to 10 mils in 0.2-mil increments. The micrometer has red and black figures. The black figures are used when taking readings on the 0–800-mil scale, while the red figures are used with the 800–1600-mil scale. With tank guns, use the black figures.

(4) Auxiliary indexes on the index arm and on the plunger indicate whether the micrometer is to be read as 0 mils or as 10 mils. A zero micrometer indication is read as 0 mils when the auxiliary indexes are matched, and as 10 mils when they are not matched.

(5) Steel shoes are screwed to the frame on two sides to serve as two sets of true bearing surfaces for the quadrant.

(6) An arrow, with the instruction LINE OF FIRE, is on each side of the frame to indicate the directions the quadrant is to be faced.

c. Test and Adjustment. Test of zero setting (end-for-end test) is accomplished as follows:

(1) Set both the index arm and micrometer at 0.

(2) Place the quadrant on the quadrant seats of the breech ring and center the bubble by elevating or depressing the gun.

(3) Turn the quadrant end for end. If the bubble recenters itself, the quadrant is in perfect adjustment at zero elevation. If the bubble does not recenter itself, try to center the bubble by turning the micrometer.

(4) If the bubble can be recentered with the micrometer, the correction is plus. It is one half the reading on the micrometer, and must be added to all settings.

(5) If the bubble cannot be recentered by turning the micrometer, set the index arm at −10 mils on the graduated arc (one notch below 0). Then center the bubble with the micrometer. In this case, the correction is minus and must be subtracted from all subsequent readings. Subtract the setting on the micrometer from 10 and divide the remainder by 2.

(6) The quadrant may still be used by applying the corrections as indicated above. If the correction exceeds 0.4 mil, send the quadrant to ordnance maintenance personnel for adjustment. Company personnel are not permitted to make any adjustment of the quadrant.

Section VI. TURRET AND GUN MANUAL AND POWER CONTROL SYSTEM

38. General

In order for the turret of the M47 tank to rotate independently of the hull, bearings are used to mount the turret on the hull. This permits a 360° traverse in either direction, both manually and by power. The 90-mm Gun, M36, is mounted on trunnions in the forward wall of the turret and can be elevated or depressed independently of the traverse of the turret. A hand elevating pump and a piston and cylinder assembly provide manual elevation or depression of the gun, while manual traverse is obtained through a traverse gear box. Power traverse and elevation are accomplished through an electrically controlled hydraulic system.

39. Data, Power Controls

Depression of gun (maximum) _____ —10°.
Elevation of gun (maximum) _____ +19°.
Power elevation rate (maximum) _____ 4° per second.
Traverse of turret_____ 360°.
Tracking rate of power traverse (maximum) _____ 175 mils per second.
Tracking rate of power traverse (minimum) _____ 0.3 mils per second.
Maximum traverse rate_____ 4 rpm (slew).
Final minimum correction_____ 0.25 mil.

40. Principles of Functioning of Turret and Gun Manual Control

a. Manual Traverse (fig. 24). Fixed to the turret is a traverse gear box which causes the turret to rotate when the manual traverse control handle is rotated in either direction. The manual traverse control handle is connected, through a mechanical gear train, to two pinion gears which engage the turret ring gear. Component parts of the gear box and their purpose or function are as follows:

    (1) *Traverse no-back.* Located in the traverse gear box, just forward of the manual traverse control handle (fig. 24), is a traverse no-back. This mechanical device permits rotation of the gears in the gear box only when the manual traverse control handle is rotated. It prevents the manual traverse control handle from spinning and injuring the gunner in the event the turret should rotate unexpectedly. Unexpected rotation of the turret may occur when the tank is moving and the gun strikes an object or when the tank makes sharp turns. The traverse no-back also serves to prevent the turret from "running ahead" when the turret is being traversed manually, especially when the tank is canted or on a slope. Still another function of the traverse no-back is to hold the turret while in manual operation, and prevent it from rotating freely on its bearings while not being traversed.

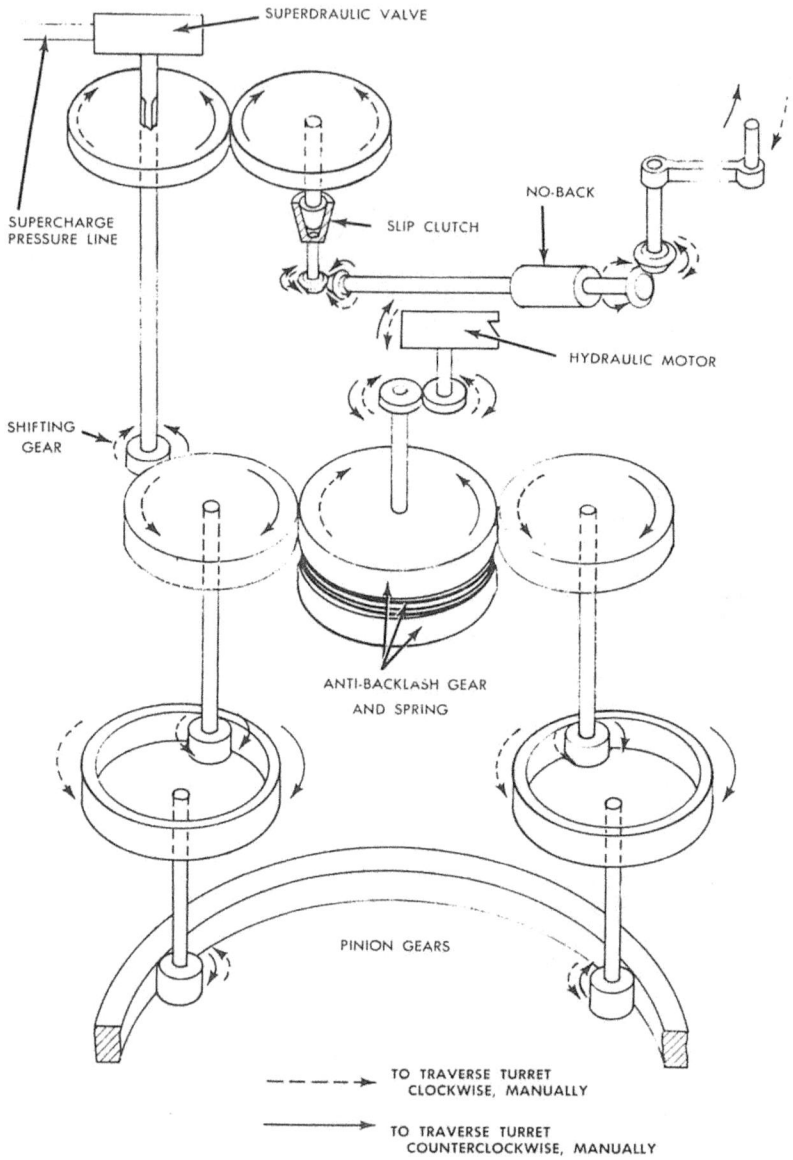

SUPERDRAULIC VALVE

SUPERCHARGE
PRESSURE LINE

NO-BACK

SLIP CLUTCH

HYDRAULIC MOTOR

SHIFTING
GEAR

ANTI-BACKLASH GEAR
AND SPRING

PINION GEARS

- - - - - → TO TRAVERSE TURRET
CLOCKWISE, MANUALLY

————→ TO TRAVERSE TURRET
COUNTERCLOCKWISE, MANUALLY

Figure 24. Traverse mechanism (manual drive).

(2) *Slip clutch.* A slip clutch has been placed in the traverse gear box to prevent damage to the gears in the event the turret rotates unexpectedly as indicated in (1) above. The clutch will slip when torque in excess of that required to operate the turret is obtained from unexpected rotation of the turret.

(3) *Shifting mechanism.* An automatic shifting mechanism in the traverse gear box functions to shift the gears in the gear box from manual to power when power is turned on and from power to manual when power is turned off, or when the dump valve is actuated while the turret power pack is in operation. (See paragraph 41b(2)(g).) Shifting from manual to power is accomplished when supercharge pressure supplied by the gear pump causes the superdraulic valve to function to disengage the gears required for manual traverse from those required for power traverse (fig. 25). Shifting from power to manual traverse can be done by turning off the turret power pack. This, in turn, will cause the supercharge pressure to drop at the superdraulic valve, allowing the gears in the traverse gear box to engage the gears required for manual traverse. Shifting can also be done by disengaging the manual traverse control handle from its locked position, thereby causing the dump valve to relieve the supercharge pressure from the superdraulic valve as indicated in paragraph 41b(2)(g).

(4) *Antibacklash gear.* To eliminate looseness or backlash in the gears of the traverse gear box, an antibacklash gear has been installed in the gear train (fig. 24). The antibacklash gear is a spring-loaded device which causes the two pinion gears to oppose each other. This causes tension to be applied to all the gears when in manual drive and to those gears being utilized when in power drive.

(5) *Pinion gears.* Two pinion gears engage the teeth on the turret ring gear to drive the turret in either direction (fig. 25). The pinion gears are connected indirectly to the turret and, when rotated, move around the ring gear which is fixed to the tank hull, consequently moving the turret.

(6) *Hydraulic motor* (par. 41b(2)(f)).

b. Manual Elevation. A hydraulic system is used to elevate or depress the gun on its trunnions in the turret. The manual elevation system consists chiefly of an elevating piston and cylinder, and a manual elevation hand pump (fig. 26).

(1) *Elevating piston and cylinder.* The elevating cylinder is connected to and moves with the gun. The piston, which is housed in the elevating cylinder, is fastened to a bracket

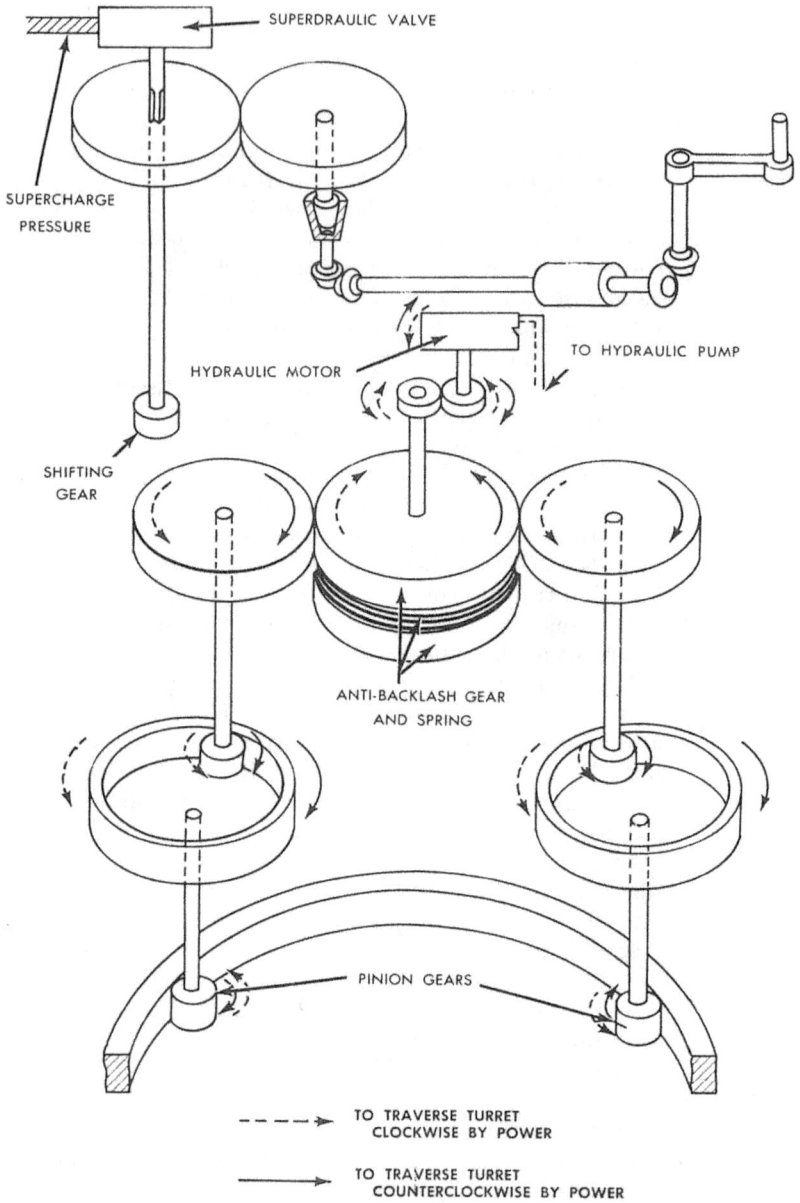

SUPERDRAULIC VALVE

SUPERCHARGE
PRESSURE

HYDRAULIC MOTOR

TO HYDRAULIC PUMP

SHIFTING
GEAR

ANTI-BACKLASH GEAR
AND SPRING

PINION GEARS

- - - - → TO TRAVERSE TURRET
CLOCKWISE BY POWER

———→ TO TRAVERSE TURRET
COUNTERCLOCKWISE BY POWER

Figure 25. Traverse mechanism (power drive).

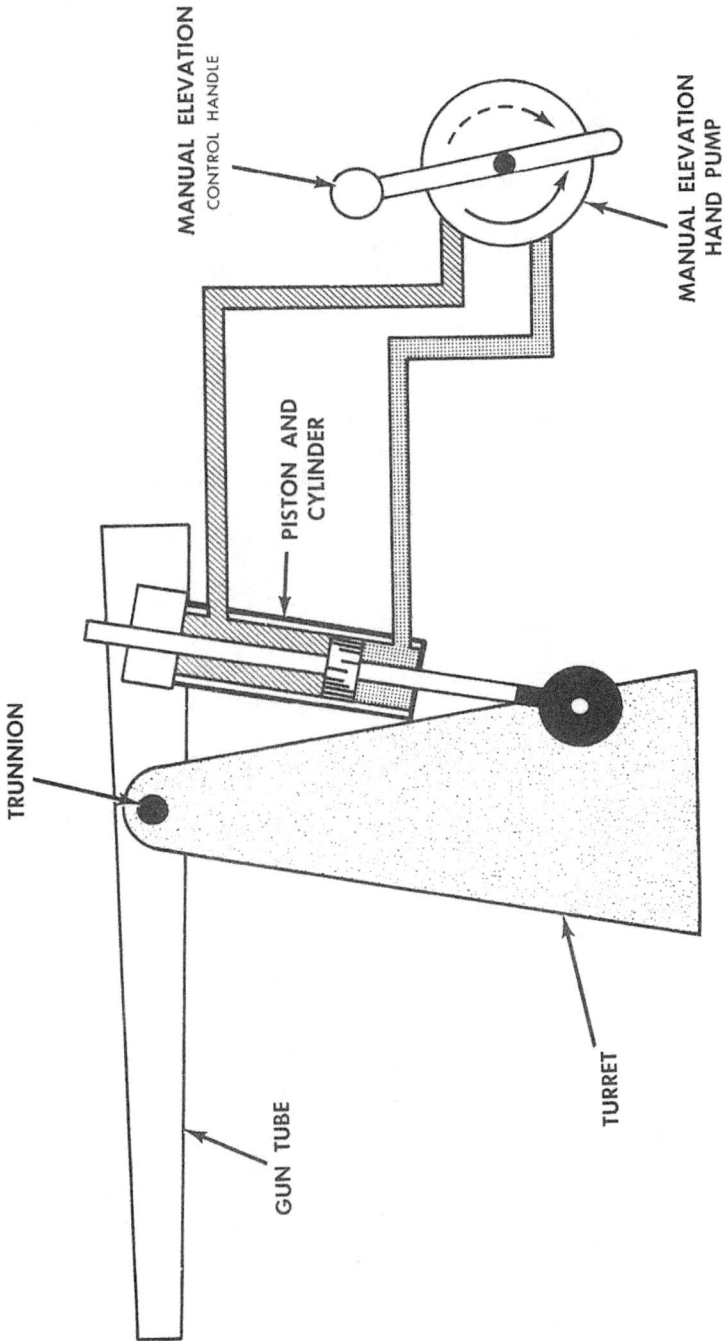

MANUAL ELEVATION
CONTROL HANDLE

MANUAL ELEVATION
HAND PUMP

PISTON AND
CYLINDER

TRUNNION

GUN TUBE

TURRET

Figure 26. Basic function of manual elevating system.

61

suspended from the turret by means of a piston rod and does not move with movement of the gun. To elevate the gun, oil is forced into the elevating cylinder below the piston and causes the cylinder to move down. At the same time, oil is allowed to escape from above the piston. To depress the gun, oil is forced into the elevating cylinder above the piston, and oil below the piston is allowed to escape.

(2) *Manual elevation hand pump.* A reversible, hand-operated hydraulic pump is used to move the oil that causes the elevating piston and cylinder to function as indicated in (1) above. To elevate the gun, the manual elevation handle is rotated in a clockwise direction, causing the manual elevation hand pump to supply oil to the elevating cylinder (fig. 27) below the piston. At the same time, the manual elevation hand pump absorbs the oil that escapes from above the piston and uses this same oil as a source of supply. To depress the gun, the manual elevation handle is rotated in a counterclock-

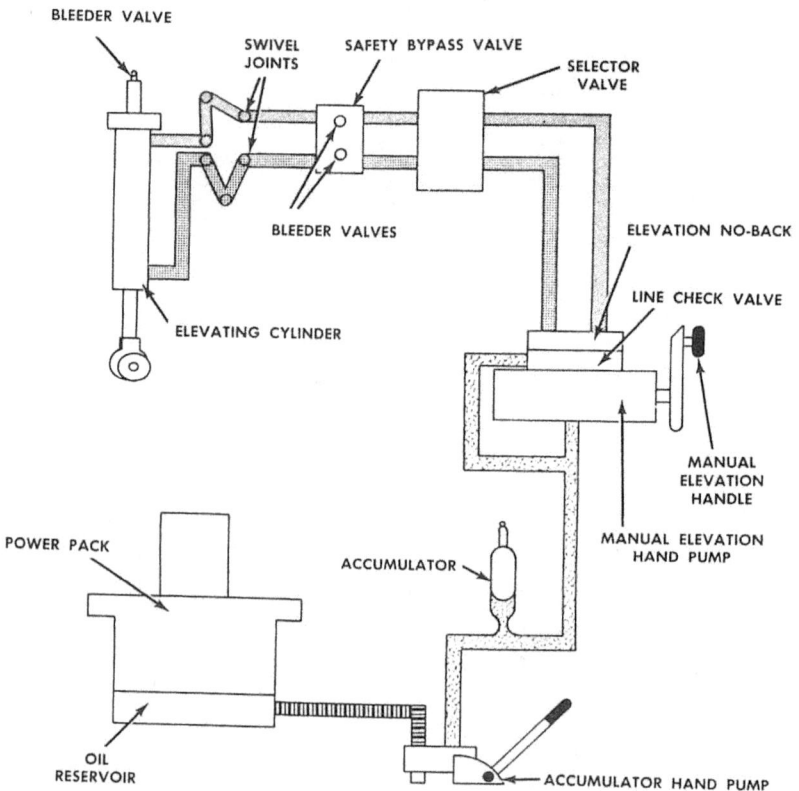

Figure 27. Manual elevation system (schematic).

wise direction, and the oil flow is reversed from that used to elevate the gun.

(3) *Elevation no-back.* The elevation no-back is a hydraulically operated device mounted on the manual elevation hand pump. The no-back is so fixed that oil flowing to, or returning from, the elevation piston and cylinder must pass through it; however, movement of the oil must come from the manual elevation hand pump (fig. 27). The elevation no-back prevents oil pressure, caused by the gun moving unexpectedly (due to movement of the tank over uneven terrain), from reaching the manual elevation hand pump, thereby preventing the manual elevation handle from spinning.

(4) *Safety bypass valve.* A safety bypass valve (fig. 27) is placed in the oil lines of the manual elevation system to prevent damage to the system in the event the gun moves unexpectedly. Since the elevation no-back prevents oil from flowing when movement is initiated at the gun, the safety bypass valve functions to relieve excessive pressures by opening just before such excessive pressure can cause damage to the system. This function allows oil to flow from one side of the piston in the elevating cylinder to the opposite side without going through the manual elevation hand pump.

(5) *Swivel joints.* Swivel joints (fig. 27) are installed in the lines of the elevation system to permit flexibility of the oil lines. This is necessary to allow oil lines that are connected to the elevating cylinder to follow the up-and-down movement of the gun.

c. Accumulator System. The accumulator system is necessary to keep the manual elevation system pressurized (fig. 27). The system consists of an accumulator hand pump, an accumulator, and a line check valve.

(1) *Accumulator hand pump.* Located to the right of the gunner's seat and at about the same level (on older model tanks, on the floor of the turret basket under the gun), is a manually operated hydraulic accumulator hand pump (fig. 27). To cause the pump to function, grasp the handle, extend it for maximum leverage, and move it back and forth. This action causes the pump to suck oil from the oil reservoir, which is fixed to the power pack, and pump it into the manual elevation system through the oil lines of the accumulator system. This is necessary to fill the manual elevation system and to keep it pressurized at all times so the gun will elevate or depress instantly when the manual elevation handle is rotated.

(2) *Accumulator.* The accumulator is installed in the accumulator system oil lines and serves to keep the manual elevation

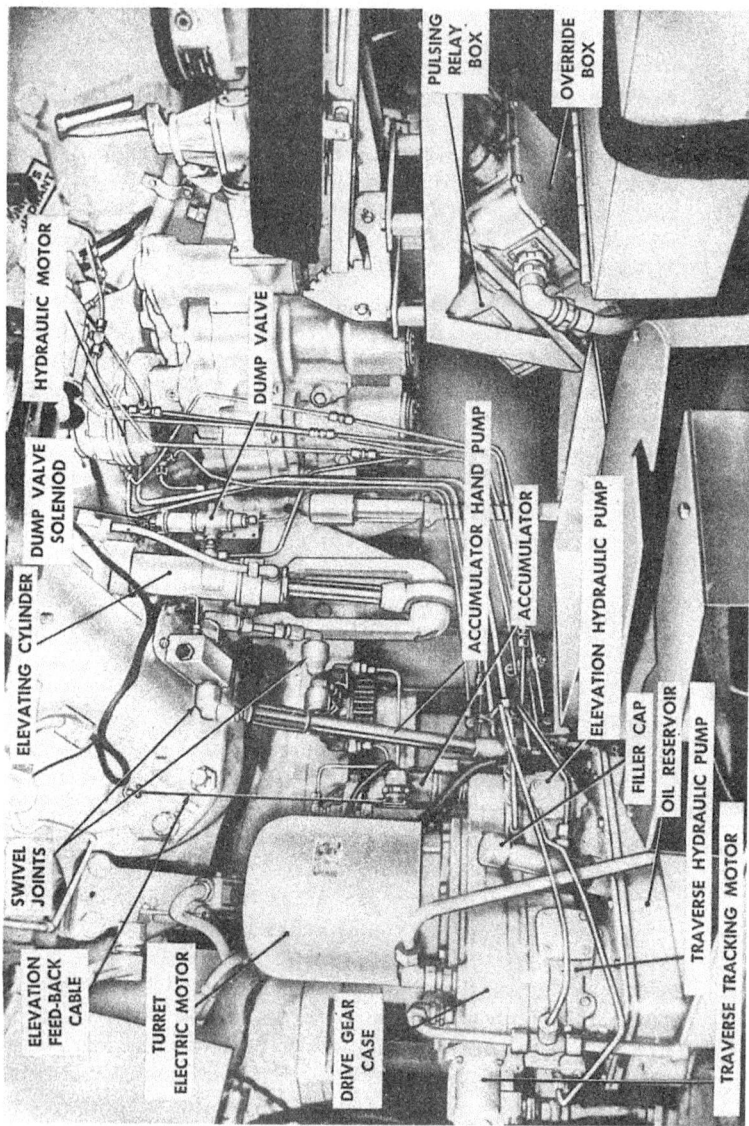

PULSING RELAY BOX

OVERRIDE BOX

HYDRAULIC MOTOR

DUMP VALVE

DUMP VALVE SOLENIOD

ELEVATING CYLINDER

ACCUMULATOR HAND PUMP

ACCUMULATOR

ELEVATION HYDRAULIC PUMP

SWIVEL JOINTS

FILLER CAP

OIL RESERVOIR

ELEVATION FEED-BACK CABLE

TURRET ELECTRIC MOTOR

TRAVERSE HYDRAULIC PUMP

DRIVE GEAR CASE

TRAVERSE TRACKING MOTOR

Figure 28. Turret traversing and gun elevating system.

64

system pressurized after the pressure has been built up by the accumulator hand pump (figs. 27 and 28). The accumulator is a metal cylinder which houses a rubber bladder or bag. Since oil cannot be compressed, the rubber bag is filled (by ordnance personnel) with nitrogen, which can be compressed. Oil that is pumped from the accumulator hand pump forces its way into the accumulator and compresses the nitrogen that is contained in the rubber bag. This compressed nitrogen is constantly trying to expand, thereby keeping the oil in the accumulator and manual elevation systems under pressure. This eliminates the need for constant pumping of the accumulator hand pump. It will be necessary to repressurize the systems from time to time by pumping the accumulator hand pump. Loss of pressure is due to small oil leaks. Leaks permit the nitrogen in the rubber bag to expand, thereby allowing the oil pressure to drop. When the accumulator hand pump is pumped, relief valves in the system prevent damage by opening when the manual elevation system has reached the proper operating pressure. The manual elevation and accumulator systems can also be pressurized by traversing the turret in either direction at a high rate of speed while operating in power. The tank crew should *never* attempt to check the nitrogen pressure in the accumulator. Depressing the core of the valve stem which is located under the cap on top of the accumulator, even though momentarily, will result in an almost instant loss of nitrogen. If it is believed that nitrogen pressure is not adequate, notify the turret artillery mechanic.

(3) *Line check valve.* The line check valve, mounted on the manual elevation hand pump, serves to prevent oil from escaping from the manual elevation system (fig. 27). Oil is allowed to flow into the manual elevation system through the line check valve when the accumulator hand pump is pumped.

d. *Bleeding of Manual Elevation System.* Air in the hydraulic system is a common malfunction. When air is present in the manual elevation system, the gun will not respond quickly to rotation of the manual elevation control handle. If pumping of the accumulator hand pump does not remedy the situation, bleed the system as follows:

(1) Elevate the gun to maximum elevation.

(2) Remove the dust screws from the two bleeder valves on top of the safety bypass valve and the dust screw from the bleeder valve on top of the elevating cylinder piston rod.

(3) Install bleeder hoses on the three bleeder valves. (Oil should be drained into a pan or returned to the oil reservoir by means

of the bleeder hoses, to reduce the amount of oil wasted during the bleeding operation.)

(4) Open all bleeder valves.

(5) Pump the accumulator hand pump until all air has been expelled from the system.

Caution: Do not turn elevation control handle during bleeding operations.

(6) Close all bleeder valves.

(7) Depress and elevate the gun to maximum positions. If there is still air in the system, repeat the procedure in (1) through (6) above.

(8) Remove bleeder hoses.

(9) Replace dust screws.

Note 1. The oil level in the reservoir must be maintained at the FULL mark on the bayonet gage.

Note 2. A manual shutoff valve is placed in the accumulator drain line just below the accumulator. This valve is provided to drain the accumulator system for maintenance purposes and should normally remain closed. Older model tanks may also be equipped with a manual shutoff valve located to the right of the gunner's power control assembly. Modifications to the manual elevation system have made this valve useless, but it must be in the closed position for manual operation.

41. Principles of Functioning of Power Elevation and Power Traverse

Power elevation and traverse are accomplished through an electrically controlled hydraulic system (fig. 29). Turning the turret motor switch on starts the turret electric motor and energizes the electrical control system. Turning the gunner's or commander's power control handle in a direction to elevate or depress the gun will cause the elevation tracking motor to operate the elevation hydraulic pump so that a pumping action of the hydraulic pump will result. Oil which is pumped by the elevation hydraulic pump will operate the same piston and cylinder assembly used in the manual elevation system to elevate or depress the gun. Rotation of the gunner's or commander's power control handle to traverse the turret in either direction will cause the traverse tracking motor to operate the traverse hydraulic pump so that a pumping action of the traverse hydraulic pump will result. Oil pumped by the traverse pump drives the hydraulic motor, which, in turn, drives the traverse mechanism to rotate the turret. (It is necessary that the main engine or the auxiliary engine be running when operating the turret power control system.)

a. Electrical Control System. The electrical control system (fig. 30) is generally known as the "pulsing relay system" and functions to control the turret power pack (fig. 31). It is composed of the gunner's and commander's power control handles, the override box, the pulsing relay box, and the tracking motors. To control the superelevation of

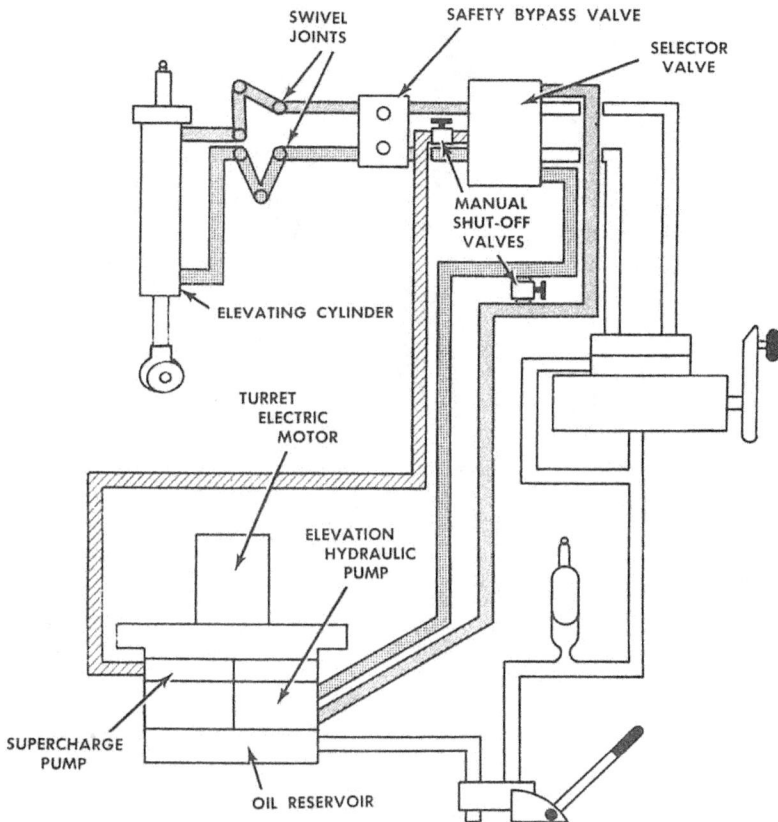

Figure 29. Power elevation system (schematic).

the gun when operating the range finder, the system also employs an inverter and a superelevation transmitter.

(1) *Gunner's power control handle (fig. 30).* The gunner's power control handle in the M47 tank may be either of two types of handles. Early model tanks are equipped with a pistol grip handle; later models have a spade-grip handle. Although different in appearance, their location and operation are practically the same. With this handle the gunner can rotate the turret in either direction at variable speeds as desired; and whether traversing or not, the gunner can elevate or depress the gun by power at variable speeds. With the turret power pack running, but with the power control handle in the neutral position for both elevation and traverse, the turret and gun will remain stationary. Moving the control handle from its neutral position to elevate or traverse will set

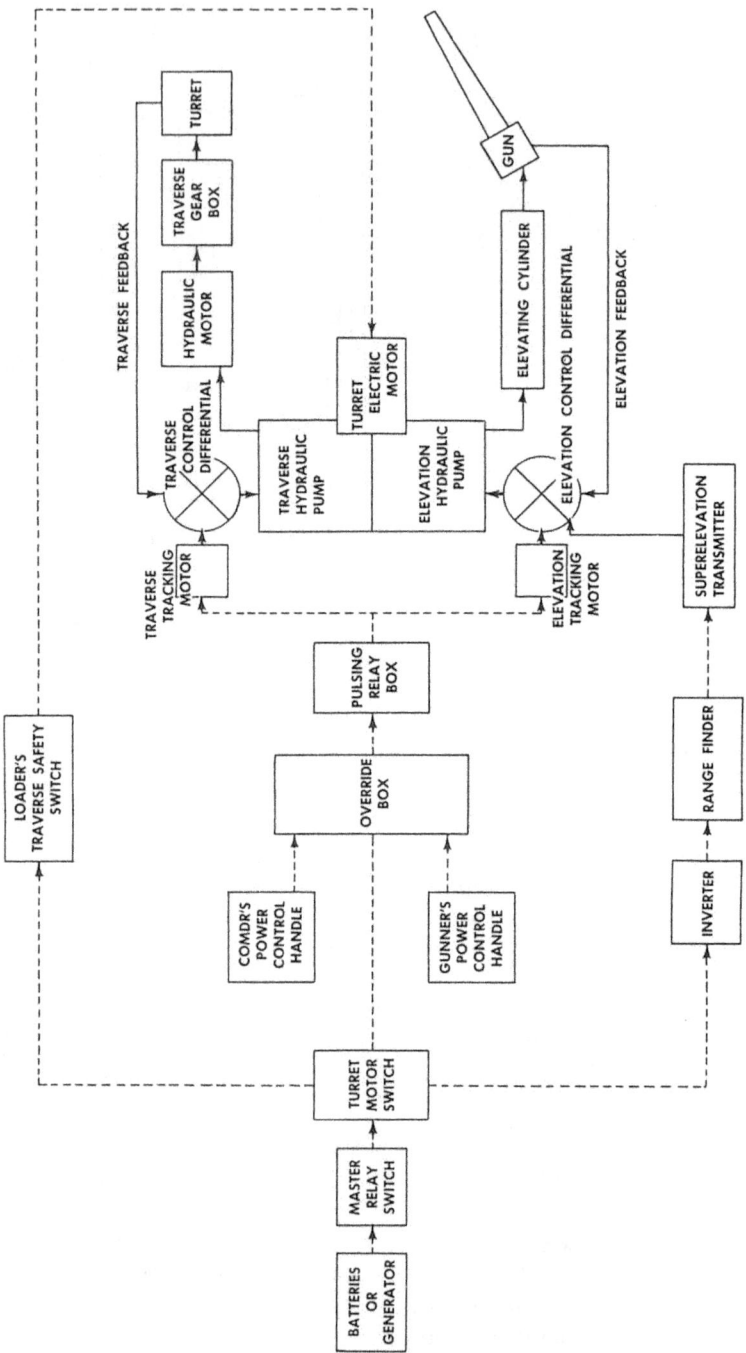

Figure 30. Block diagram of turret power control system.

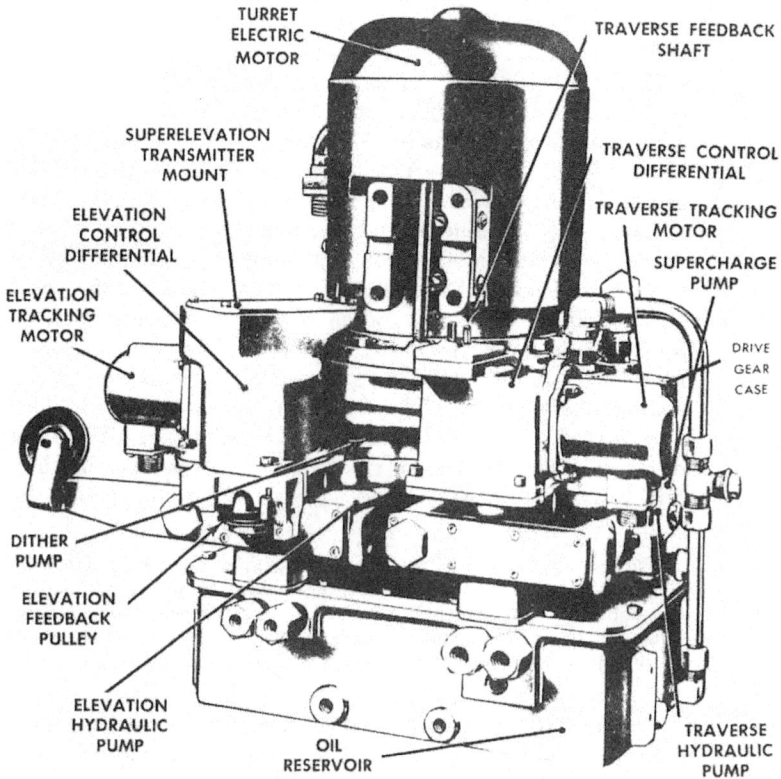

TURRET
ELECTRIC
MOTOR

TRAVERSE FEEDBACK
SHAFT

SUPERELEVATION
TRANSMITTER
MOUNT

TRAVERSE CONTROL
DIFFERENTIAL

ELEVATION
CONTROL
DIFFERENTIAL

TRAVERSE TRACKING
MOTOR

ELEVATION
TRACKING
MOTOR

SUPERCHARGE
PUMP

DRIVE
GEAR
CASE

DITHER
PUMP

ELEVATION
FEEDBACK
PULLEY

ELEVATION
HYDRAULIC
PUMP

OIL
RESERVOIR

TRAVERSE
HYDRAULIC
PUMP

Figure 31. Power pack (rear view).

up a flow of electricity. The direction of flow can be controlled by the direction the handle is turned. The amount of current is determined by the distance the handle is moved from neutral. This action allows electrical current to flow through the override box to the pulsing relay box and finally to the tracking motors. The speed of either tracking motor is determined by the amount of current supplied by the control handle. Turning the power control handle to its maximum displacement in any of the four directions will cause "slew" (maximum traverse speed) switches to be actuated, and the tracking motor will rotate at its maximum speed. When the slew switches are actuated, variable speeds cannot be obtained. Returning the power control handle to the neutral position will stop rotation of the tracking motor.

(2) *Commander's power control handle.* A pistol-grip type handle is used by the commander to traverse the turret in power. The function of the commander's power control

handle is like that of the gunner's except that the commander's control handle employs an override lever. The override lever, when depressed, will cause the gunner's elevation and traverse controls to become inactive, and full elevation and traverse control is assumed by the tank commander. This is accomplished as in (3) below.

(3) *Override box.* The override box (fig. 30), located under the gunner's seat, serves as a junction box for the pulsing relay system. Both gunner's and commander's control of the tracking motor passes through the override box, which is equipped with switches that are normally positioned for gunner operation. However, when the tank comander presses his override lever, the switches in the override control box automatically disengage the gunner's circuit and engage the commander's circuit. The comander has control of the gun and the turret until he releases his override lever. When the override lever is released, the switches in the override box will disengage the commander's circuit and again engage the gunner's circuit.

> *Note.* With the tank in power and the gunner traversing manually, the tank commander can override the gunner by pressing the commander's override lever and rotating the commander's power control handle. If the commander's override lever is released before the turret has been brought to a standstill, damage to the traversing mechanism will result.

Correction: Until a permanent modification can be made, line No. 623, top single line connected to the rear of the commander's power traversing and elevating control, is not being connected by the manufacturer. If line No. 623 is connected, disconnect it and tape both connections. Tape line to the electrical harness to prevent damage.

(4) *Pulsing relay box.* The pulsing relay box completes the electrical circuit from the override box to the tracking motors (fig. 30). It serves to convert the steady flow of low-voltage current coming from either the gunner's or commander's power control handle to an intermittent pulse or "on-and-off" flow of 24-volt current. The amount of current supplied the pulsing relay box determines the frequency of the pulses, thereby controlling the speed of the elevation or traverse tracking motor, depending on the position of the control handle. The purpose of this relay is to allow a full 24 volts to reach the tracking motor with each pulse in order to give the motor more power when operating at its minimum speed of six revolutions per minute.

(5) *Elevation tracking motor.* The elevation tracking motor is a reversible, variable-speed electric motor which is secured

to the turret power pack (fig. 31). Its purpose is to regulate the amount of oil pumped by the elevation hydraulic pump.

(6) *Traverse tracking motor (fig. 31).* The traverse tracking motor is identical to the elevation tracking motor, except that the traverse tracking motor regulates the amount of oil pumped by the traverse hydraulic pump.

(7) *Inverter.* The inverter is used only in conjunction with the range finder and superelevation transmitter. It is located in the turret behind the radio, and serves to change the 24-volt dc current supplied by the tank electrical system to a 110-volt, 400-cycle ac current (fig. 30). The servo mechanisms in the range finder and the superelevation transmitter are designed to operate on 110 volts ac; therefore the inverter is necessary. (The reticle lighting system in the range finder uses 24 volts dc.)

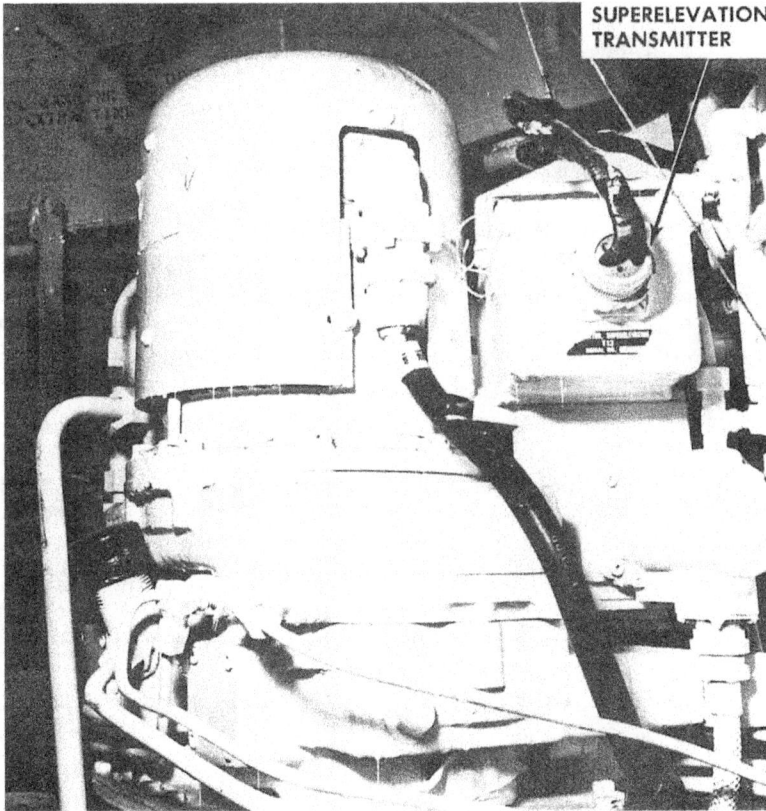

Figure 32. Position of superelevation transmitter on power pack.

(8) *Superelevation transmitter.* (Superelevation is the *angle above the line of sight* that the gun must be elevated to obtain a target hit at a given range.) The superelevation transmitter (fig. 32) is fixed to the turret power pack and functions to control the pumping action of the elevation tracking motor. The major difference between the elevation tracking motor and the superelevation transmitter is that the superelevation transmitter is put into action as the ranging knob on the range finder is being rotated, while operation of the elevation tracking motor is accomplished independently of the superelevation transmitter by moving the power control handle from its neutral position. Therefore, rotating the ranging knob on the range finder to increase range will cause an electrical signal to be sent from the range finder to the superelevation transmitter. The superelevation transmitter converts this signal from electricity to motion and causes the elevation hydraulic pump (fig. 30) to function to elevate the gun automatically. Rotation of the ranging knob to decrease range will cause the gun to be depressed automatically. This eliminates the need for transferring the range from the range finder to a superelevation knob, and serves to keep the ranging reticle at about the same elevation while ranging. (Gunners must be cautioned that the ranging knob on the range finder should not be rotated while the tank commander is attempting to lay and fire the gun.)

b. Hydraulic Elevation and Traverse System (Power). Power to elevate the gun and drive the turret in traverse is furnished by the turret power pack. The turret electric motor, independent of the pulsing relay system, drives the power pack, which furnishes the necessary oil pressures to shift from manual to power elevation, to shift from manual to power traverse, to drive the hydraulic motor, and to cause the elevating cylinder assembly to function (fig. 33). The elevation and traverse tracking motors, which are controlled by the gunner's and commander's power control handles, control the oil pressures that cause the elevating cylinder to function and the hydraulic motor to rotate. This is accomplished when either tracking motor, through its control differential, moves the eccentric shaft. The eccentric shaft, in turn, causes the respective pilot valve to put the elevation on traverse hydraulic pumps into action (fig. 34).

(1) *Hydraulic elevation system.* The system consists chiefly of a turret electric motor, a gear or supercharge pump, an elevation hydraulic pump, and a selector valve. It utilizes the same elevating cylinder and piston assembly as the manual elevation system (fig. 29). The system also contains an ele-

vation control differential and a cable-type elevation feedback.

(a) *Turret electric motor.* Turning the turret motor switch (fig. 30) to the ON position will start the motor, providing the loader's traverse safety switch (fig. 30) is also on. Turning the loader's traverse safety switch off will stop the motor, as will the turret motor switch. The turret electric motor is mounted on top of the power pack with the drive end down (fig. 31). The motor runs at a constant speed in one direction only and drives two drive shafts. The drive shafts, through the drive gears, operate the supercharge or gear pump, the elevation and traverse hydraulic pumps, and the dither pump.

(b) *Supercharge or gear pump.* The supercharge pump is a constant displacement pump that fills the entire hydraulic system when the turret electric motor is started. It also supplies the supercharge pressure that causes the selector valve to function ((d) below), and the superdraulic valve to be actuated as in (g) below, and furnishes the necessary operating pressure for the elevation and traverse hydraulic pumps (fig. 33).

(c) *Elevation hydraulic pump.* Located in the turret power pack (fig. 31), the pump is driven through the drive gear and shaft by the turret electric motor. The pump rotates at a constant speed as long as the turret electric motor is in operation (fig. 34). However, unless the slide in the pump is moved from its neutral position, it will not pump oil. To move the slide block from its neutral position, the elevation tracking motor turns a control differential; the control differential moves the eccentric shaft from its neutral position and causes the pilot valve to function. The pilot valve, through supercharge oil pressure, moves the slide block to start pumping action of the elevation hydraulic pump.

(d) *Selector valve.* The selector valve is located just above the turret power pack and serves as the shifting mechanism for the manual and power elevation systems (fig. 29). With turret power off, the ports in the selector valve are so positioned that oil may be pumped through them by the manual elevation hand pump. This causes the elevating cylinder and piston assembly to function. When turret power is turned on, supercharge pressure furnished by the supercharge pump will cause the ports in the selector valve to shift automatically and position the ports so that the manual elevation hand pump is blocked

A. HYDRAULIC PUMP IN NEUTRAL POSITION.

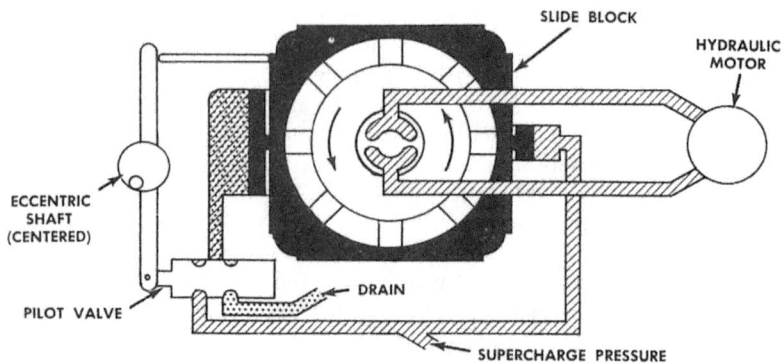

SLIDE BLOCK

HYDRAULIC MOTOR

ECCENTRIC SHAFT (CENTERED)

PILOT VALVE

DRAIN

SUPERCHARGE PRESSURE

B. SLIDE BLOCK MOVED TO OBTAIN RIGHT TRAVERSE.

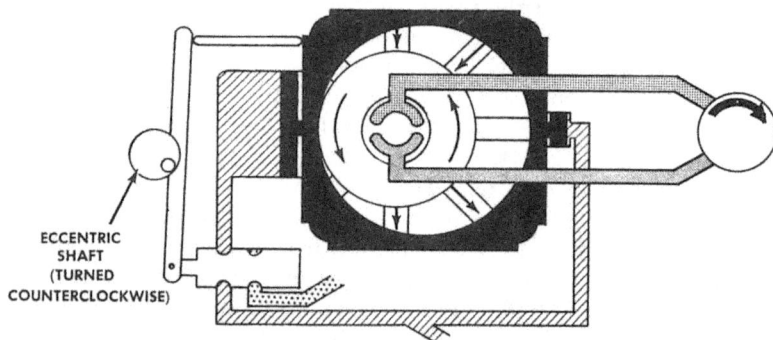

ECCENTRIC SHAFT (TURNED COUNTERCLOCKWISE)

C. SLIDE BLOCK MOVED TO OBTAIN LEFT TRAVERSE.

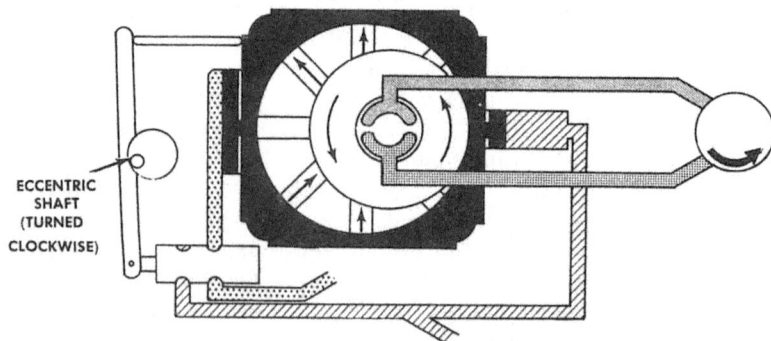

ECCENTRIC SHAFT (TURNED CLOCKWISE)

Figure 34.   Function of hydraulic pump.

out. Oil may then be pumped through the selector valve by the elevation hydraulic pump to cause the gun to be elevated or depressed through action of the elevating cylinder. The selector valve will automatically shift back to manual when turret power is turned off. (The gun cannot be elevated manually while turret power is turned on.)

(e) *Elevating cylinder and piston.* The function of the elevating cylinder and piston when used for power elevation is the same as outlined in paragraph 40b (1) for manual elevation except that the operating oil pressure is furnished by the elevation hydraulic pump.

(f) *Manual shutoff valves.* In the event power elevation fails and manual elevation is desired while traversing in power, two manual shutoff valves are provided to block out power elevation. These valves should be positioned as follows:

1. Located on top of the selector valve is a manual shutoff valve which controls supercharge pressure to the selector valve. The shutoff valve normally should be open. To prevent the selector valve from shifting from manual to power elevation when turret power is turned on, close this valve while turret power is off.

2. Located below the selector valve and serving as a connector between two hydraulic lines leading from the elevation hydraulic pump is a second manual shutoff valve which is normally closed for power elevation. However, when closing the manual shutoff valve discussed in *1* above, it is necessary that this valve be opened. (When power elevation is blocked out, the superelevation transmitter cannot function to set in superelevation automatically while ranging.)

(g) *Elevation control differential and feedback.* The elevation control differential is mounted on the turret power pack (fig. 31), and the elevation feedback cable is attached to the gun so the cable can follow the up-and-down movement of the gun (fig. 30) and transmit this movement to the control differential. Movement of the gunner's or commander's power control handle to elevate or depress the gun causes the tracking motor to spin at a speed which is determined by the position of the handle. The tracking motor turns the control differential; and since the gun is in a stationary position, the differential turns the eccentric shaft, which, in turn, moves the pilot valve. This action causes the slide block to be moved so that the hydraulic pump is on stroke and pumps oil to elevate or depress the gun. As the gun starts to move, the elevation feedback cranks into the control differential a rotation opposite to that supplied by the elevation tracking motor, so that the tracking motor must continue to run to hold the elevation hydraulic pump on stroke. When the operator's handle is returned to the neutral position, the elevation tracking motor stops, and the feedback works through the control differential to return the eccentric shaft to the neutral position and stop movement of the gun. The elevation feedback also functions to pick up any irregularities in the movement of the gun. This function keeps the movement of the gun smooth or constant by transmitting irregularities to the control differential, which, in turn, will cause the eccentric shaft to be rotated. This causes the hydraulic pump to increase or decrease its pumping action until the gun is again moving without irregularity, at which time the eccentric shaft will assume its former position.

(2) *Hydraulic traverse system.* Power to drive the turret in traverse is furnished by the turret power pack. The same turret electric motor that drives the elevation system also drives the power pack. The power pack, in turn, furnishes the necessary oil pressures to shift the traversing mechanism from manual to power and to drive the hydraulic motor (fig. 35). The traverse tracking motor causes the traverse hydraulic pump to be put into action in exactly the same manner as the elevation hydraulic pump in (1) (*c*) above. The traverse hydraulic system consists of the following major components, some of which are also used in the power elevation system; the turret electric motor, a gear or supercharge pump, a traverse hydraulic pump, an oil reservoir, a dump valve, a superdraulic motor, and a traverse control differential to include the traverse feedback (fig. 30). (The power traverse system is self bleeding.)

(*a*) *Turret electric motor.* In addition to the functions of the turret electric motor (fig. 31) in (1) (*a*) above, the motor also serves to drive the traverse hydraulic pump and the dither pump for the traversing system.

(*b*) *Supercharge or gear pump* ((1) (*b*) above). The supercharge pump also supplies the supercharge pressure that causes the superdraulic valve to function as outlined in paragraph 40*a*(3), and to furnish the necessary operating pressure for the traverse hydraulic pump.

(*c*) *Traverse hydraulic pump.* The function of the traverse hydraulic pump is exactly like that of the elevation hydraulic pump except that it is controlled by the traverse tracking motor.

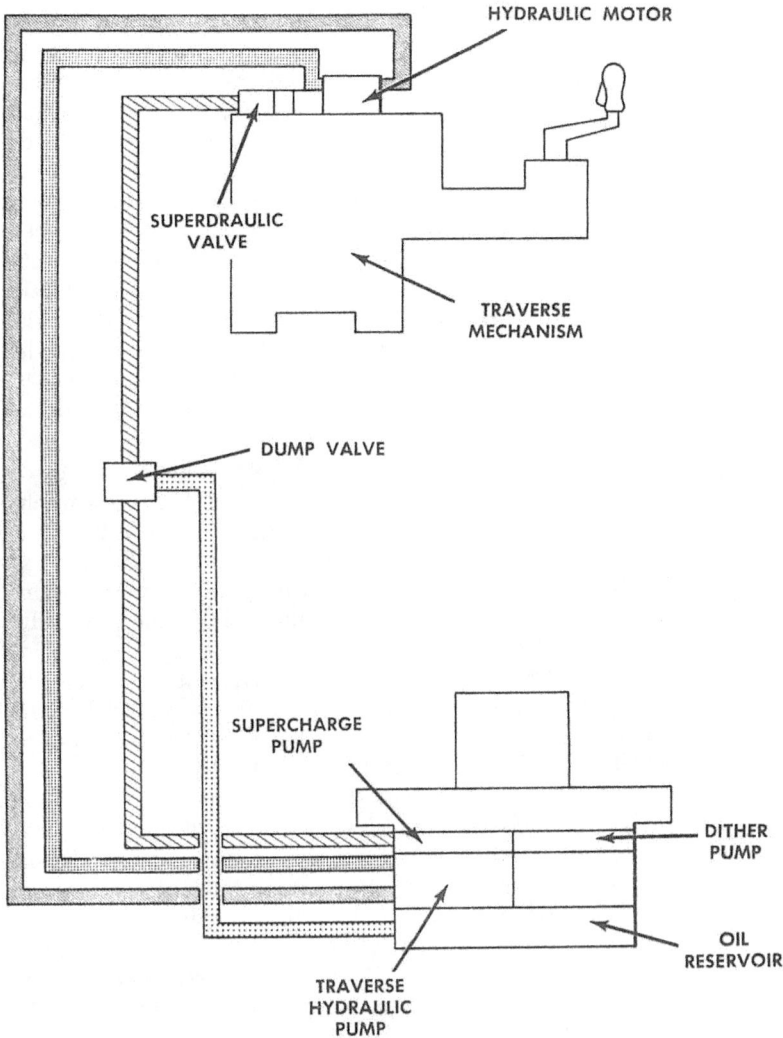

Figure 35. Power traverse system (schematic).

(d) *Dither pump.* The dither pump is located in the turret power pack (fig. 31). Its purpose is two-fold. It keeps the oil (in the traverse system only) pulsing or fluctuating so the system's working parts will not freeze because of vibrations during periods of idleness while the turret electric motor is running. The pump also helps start the turret moving gradually the instant either of the two power control handles is turned from its neutral position. This prevents an initial surge from occurring because of the working parts being stuck.

(e) *Oil reservoir.* Located at the bottom of the turret power pack, the oil reservoir (figs. 27 and 28) serves as a replenisher for the entire hydraulic system, including manual elevation. Attached to the oil reservoir is an oil filler neck used to check the oil level and fill the reservoir to the proper level. To check the oil level, remove the cap from the oil filler neck and withdraw the bayonet gage that is fastened to the cap. Wipe the bayonet gage dry and reinsert it into the oil filler neck. Make certain the cap is seated squarely on the filler neck but not screwed down. Withdraw the gage again and note the oil level. If it does not reach the full mark, add oil. The capacity of the reservoir is $1\frac{1}{2}$ gallons. (See LO 9–7010 for proper type and grade of oil.)

(f) *Hydraulic motor.* The hydraulic motor is located on top of the traverse gear box and serves to drive the traverse mechanism when in power operation (fig. 35). The motor is driven by hydraulic oil which is supplied by the traverse hydraulic pump. The speed of the hydraulic motor is controlled by the amount of oil supplied by the hydraulic pump.

(g) *Dump valve.* A dump valve (fig. 35), which allows rapid shifting from power to manual traverse, is utilized in the supercharge line that leads to the superdraulic valve or shifting mechanism (par. 40a(3)). By removing the manual traverse control handle from its locked position, a microswitch is actuated. This causes a solenoid, which is mounted on the dump valve, to function to shift the oil passages or ports in the dump valve so that the supercharge pressure required to actuate the superdraulic valve is dumped or routed back to the oil reservoir, allowing the superdraulic valve (fig. 25) to shift the traverse mechanism from power to manual control almost instantly. This device makes manual control possible while the power pack is in operation. To shift back to power control, it is necessary only to return the manual traverse control handle to its locked position. The dump valve has an advantage over the turret motor switch in that the operator does not have to wait until the turret electric motor has stopped running before he can assume manual control. (Actuating the commander's override lever when in power operation will take both manual and power traverse as well as power elevation control away from the gunner.)

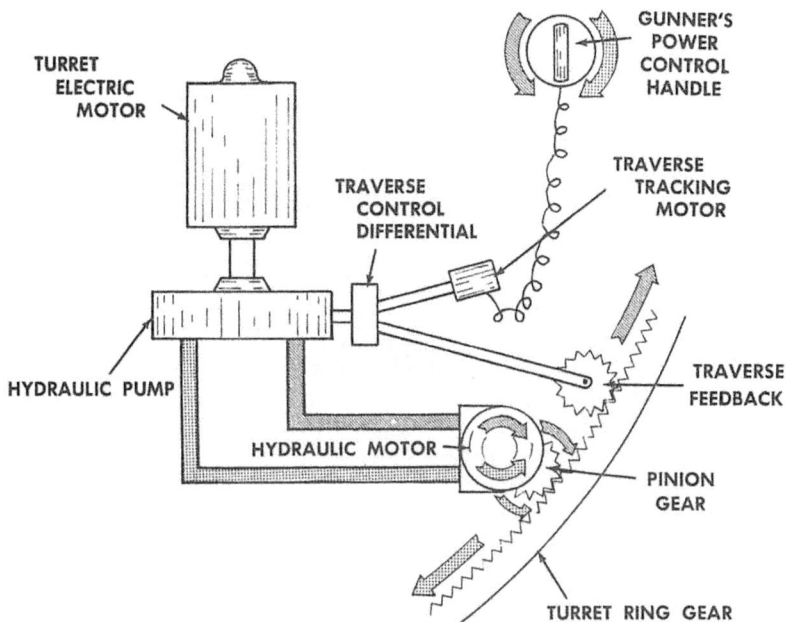

Figure 36. Basic principles of the power traverse system.

(*h*) *Traverse control differential and feedback.* The traverse
control differential is mounted on the traverse side of the
turret power pack (fig. 31), and the traverse feedback gear
is mounted on the turret so that the gear teeth engage the
teeth of the turret ring gear (fig. 36). The feedback gear
is connected to the traverse control differential by means
of a mechanical linkage. Movement of the gunner's or
commander's power control handle causes the traverse
tracking motor to spin at a rate depending on the position
of the control handle. The traverse tracking motor turns
the traverse control differential; and because the turret is
stopped, the differential turns the eccentric shaft which, in
turn, moves the pilot valve. This action causes the slide
block to be moved so that the traverse hydraulic pump is
put on stroke and pumps oil to traverse the turret. As the
turret starts to move, the traverse feedback cranks into the
control differential a rotation opposite to that supplied by
the traverse tracking motor, so that the tracking motor
must continue to run to hold the hydraulic pump on stroke.
When the operator's control handle is returned to the neutral position, the traverse tracking motor stops, and the

SQUARE

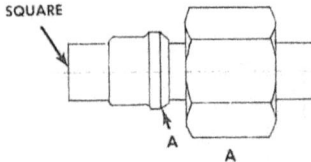

1—SLIDE NUT AND THEN THE SLEEVE ON TUBE. BE
SURE SLEEVE IS NOT ON BACKWARDS. HEAD OF
SLEEVE "A" MUST BE TOWARDS NUT.

LUBRICATE

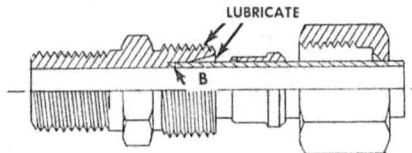

B

2—INSERT TUBE INTO FITTING. BE SURE TUBE IS
BOTTOMED ON FITTING SHOULDER AT POINT "B."
LUBRICATE WITH OIL.

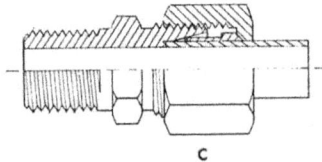

C

3—TURN NUT SLOWLY WITH WRENCH WHILE TURNING
TUBE OR CONNECTOR WITH OTHER HAND.
WHEN SLEEVE GRIPS THE TUBE OR CONNECTOR,
THAT IS, WHEN THE TUBE OR CONNECTOR CAN NO
LONGER BE ROTATED BY HAND—STOP AND NOTE
POSITION OF WRENCH.
NOW TURN NUT 1¼ TO 1½ ADDITIONAL TURNS.
FITTING WILL NOW FEEL TIGHT.
USE ONE (1) TO ONE AND ONE QUARTER (1¼)
TURNS FOR SOFT TUBING SUCH AS FULLY AN-
NEALED COPPER, ALUMINUM, ETC.

4—REASSEMBLY—TURN NUT UNTIL THE TORQUE BE-
GINS TO RISE SHARPLY.
APPLY ⅛ TO ¼ TURN FROM THAT POINT.

*Figure 37. Assembly instructions for
flareless tube fittings.*

feedback works through the control differential to return
the eccentric shaft to the neutral position and stop the
turret's rotation. The traverse feedback also picks up any
irregularities in the rotation of the turret, keeping the rotation smooth or constant by transmitting these irregularities
to the control differential, which, in turn, causes the eccentric shaft to be rotated. This causes the hydraulic pump
to increase or decrease its pumping action until the turret
is again rotating without irregularity, at which time the
eccentric shaft will assume its former position.

80

GASKET OR "O" RING

BOSS

C

TIGHTENED POSITION

PROCEDURE FOR INSTALLATION OF NON-POSITION-ING TYPE FITTING. OPERATING PRESSURES UP TO 3000 P.S.I.

1—USE GOOD LUBRICANT ON GASKET AND INSTALL IT.

2—SCREW THE FITTING ASSEMBLY INTO BOSS UNTIL THE SHOULDER OF THE HEX CONTACTS THE BOSS TIGHTLY.

LEATHER RING

"O" RING

B

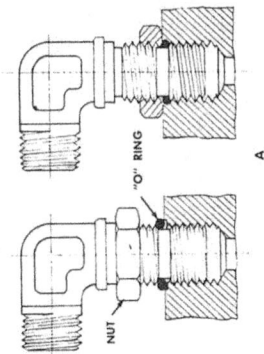

PROCEDURE FOR INSTALLATION OF UNIVERSAL FIT-TINGS FOR USE IN HYDRAULIC AND PNEUMATIC SYSTEMS OVER 1500 P.S.I.

1—INSTALL NUT SO THAT NO PART OF THE NUT COVERS THE FITTING "O" RING GROOVE. RECESS OF THE NUT SHOULD FACE THE "O" RING GROOVE.

2—INSTALL THE LEATHER BACK-UP RING IN THE GROOVE.

3—USE GOOD LUBRICANT ON "O" RING AND IN-STALL IT.

4—SCREW THE FITTING ASSEMBLY INTO THE BOSS UNTIL THE "O" RING JUST CONTACTS THE FACE OF THE BOSS.

5—POSITION FITTING BY "SCREWING IN" BY NOT MORE THAN 360°.

6—TIGHTEN LOCK NUT UNTIL IT CONTACTS THE BOSS FACE.

7—ASSEMBLE TUBE TO FITTING.

8—RE-TIGHTEN LOCK NUT AGAINST BOSS.

NOTE—CONNECTOR MUST BE HELD WITH A WRENCH WHEN ANY TIGHTENING OPERATION IS PERFORMED.

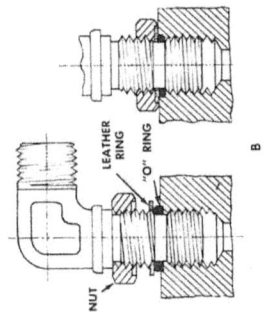

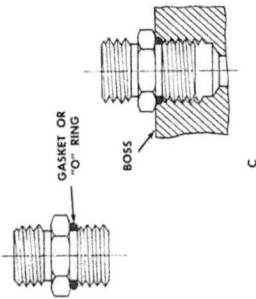

NUT

"O" RING

A

NUT

PROCEDURE FOR INSTALLATION OF UNIVERSAL FITTINGS FOR USE IN HYDRAULIC AND PNEU-MATIC SYSTEMS—1500 P.S.I. AND UNDER.

1—INSTALL NUT SO THAT NO PART OF THE NUT COVERS THE FITTING "O" RING GROOVE.

2—USE GOOD LUBRICANT ON "O" RING AND IN-STALL IT.

3—SCREW FITTING ASSEMBLY INTO BOSS UNTIL THE "O" RING JUST CONTACTS THE FACE OF THE BOSS.

4—TO POSITION FITTING "UNSCREW IT" NOT MORE THAN 360°

5—TIGHTEN LOCK NUT UNTIL IT CONTACTS BOSS FACE.

6—ASSEMBLE TUBE TO FITTING.

7—RE-TIGHTEN LOCK NUT AGAINST BOSS.

NOTE—CONNECTOR MUST BE HELD WITH WRENCH WHEN TIGHTENING OPERATION IS PERFORMED.

Figure 38.   Assembly and installation procedure for universal, straight-threaded connectors and bulkhead flareless tube fittings.

42. Assembly Instructions for Flareless Tube Fittings

Oil leaks which may develop in the hydraulic system from time to time can be corrected by the tank crew member providing the steps outlined in figures 37 and 38 are followed to the letter.

Note. When slow leaks develop at a tube fitting, tighten the nut *only* until the leak stops. *Never tighten the nut as tight as it will go.*

CHAPTER 3

CREW DRILL, SERVICE OF THE PIECE, AND STOWAGE

Section I. CREW COMPOSITION AND FORMATIONS

43. General

This chapter is for the guidance of platoon leaders and tank commanders in training crew members with a view to attaining efficient teamwork in the crew operation of the Tank, 90-mm Gun, M47. It is emphasized that the drills described in this chapter are for the development of crew teamwork in the fighting operation of the tank and that the ultimate goal to be obtained is successful operation of the tank on the battlefield.

44. Crew Composition

The tank crew of the Tank, 90-mm Gun, M47, consists of five members designated as follows:

```
Tank commander_____ TANK COMMANDER
Gunner_____ GUNNER
Bow gunner (assistant driver)_____ BOG
Driver_____ DRIVER
Loader_____ LOADER
```

45. Formations

a. Dismounted Posts. The crew forms in one rank with the tank commander 2 yards in front of the right track. The gunner, bow gunner, driver, and loader take posts on line with, and to the left of, the tank commander at close interval.

b. Mounted Posts. The crew forms mounted as follows:

(1) *Tank commander.* In the turret, standing on the tank commander's platform or seated on the tank commander's seat.

(2) *Gunner.* In the gunner's seat to the right of the tank gun and in front of the tank commander.

(3) *Bow gunner.* In the bow gunner's seat in the right front of the hull.

(4) *Driver.* In the driver's seat in the left front of the hull.

(5) *Loader.* On the left side of the tank gun, standing on the turret floor or seated on the loader's seat at the left rear of the turret.

Section II. CREW CONTROL

46. Operation of Interphone and Radio

The tank interphone system is used for voice communication among members of the tank crew and for communication with individuals outside the tank through the external interphone. The tank radio set is used for communication with other tanks and with other units. The interphone is a part of the vehicular radio set. The equipment is designed so that operation of the interphone system will override received or transmitted signals, but will not cut transmitted signals off the air. The crew must be proficient in the operation of the interphone system if they are to obtain its maximum value in combat. Proficiency in the operation of the interphone system is gained only by continued practice.

47. Control Box Positions

Interphone control box positions are as follows:

a. The tank commander and gunner plug into a single box located on the right wall of the turret.

b. The loader plugs into a control box on the left wall of the turret.

c. The driver and bow gunner plug into a single control box located between these two crew members.

48. Modes of Operation, Interphone Control Boxes

a. General. When power has been supplied to Set 1, Set 2, and the auxiliary receiver, and after squelch adjustments have been made, the following modes of operation are possible at each interphone control box:

(1) Monitoring of Set 1, Set 2, and the auxiliary receiver.

(2) Push-to-talk operation of Set 1 or Set 2.

(3) Interphone facilities between interphone boxes.

b. Monitoring. Monitoring received signals is accomplished by placing the selector switch pointer of the interphone control box in the center position. This position permits monitoring Set 1, Set 2, and interphone in the Radio Set AN/GRC–4, –6, and –8. It also permits monitoring the auxiliary receiver in the Radio Set AN/GRC–3, –5, and –7.

c. Interphone Operation. Interphone reception is possible with the selector switch in any position. To communicate with a crew member at any interphone box, press the interphone turn to LOCK button on the chest set and talk into the microphone. In an emergency, any crew member can override a radio conversation without waiting for the sending party to stop talking.

d. Radio Operation of Sets 1 and 2.

(1) For push-to-talk operation of Set 1, turn the selector switch pointer to the left-hand position, press the HOLD ON button

and the RADIO button on the chest set, place the RADIO
TRANS switch on the control box in the TRANS position,
and talk into the microphone. Release the chest set switches
to listen. If the auxiliary receiver interferes with operation
of Set 1, turn the receiver VOLUME control to the OFF
position.

(2) For push-to-talk operation of Set 2, turn the selector switch
pointer to the right position, press the RADIO button on
the chest set, place the RADIO TRANS switch on the control
box in the TRANS position, and talk into the microphone.
Release the chest set switches to listen.

e. Set 2. When Set 2 is used, the loader is designated as monitor-
operator.

49. Radio Check

Inspection of communication equipment will be performed as
prescribed on DA Form 11–238.

50. Checking Interphone Equipment

It is the duty of each crew member to check his interphone equipment. He should see that it is complete, in good working order, clean,
and properly maintained. Any difficulties should be reported to the
tank commander.

51. Use of Definite Terminology

Terminology prescribed for tank commanders in controlling their
crews, as set forth in paragraph 52, must always be used. Failure
to use standard, specific interphone language causes misunderstanding
and disorder. Adherence by all crew members to this standard language is essential to efficient operation of the tank.

52. Interphone Language

a. Terms.

Tank Commander	TANK COMMANDER
Driver	DRIVER
Gunner	GUNNER
Loader	LOADER
Bow gunner	BOG
Any tank	TANK
Any unarmored vehicle	TRUCK
Any antitank gun or artillery piece	ANTITANK
Infantry	TROOPS
Machinegun	MACHINEGUN
Airplane	PLANE
Any other target	Briefest descriptive word or phrase

b. Commands for Movement of Tank.

To move forward_____	DRIVER MOVE OUT
To halt_____	DRIVER STOP
To reverse_____	DRIVER REVERSE
To increase speed_____	DRIVER SPEED UP
To decrease speed_____	DRIVER SLOW DOWN
To turn right (left)_____	DRIVER RIGHT (LEFT)—STEADY—ON
To turn right (left) 180°_____	DRIVER RIGHT (LEFT) ABOUT—STEADY—ON
To move toward a terrain feature or reference point, tank being headed in proper direction _____	DRIVER MARCH ON WHITE HOUSE (HILL, DEAD TREE, ETC.)
To follow in column_____	DRIVER FOLLOW THAT TANK (DRIVER FOLLOW TANK B–9)
To follow road or trail to the right (left)_____	DRIVER RIGHT (LEFT) ON ROAD (TRAIL)
To start engine_____	DRIVER TURN IT OVER
To stop engine_____	DRIVER CUT ENGINE
To proceed to a specific transmission range_____	DRIVER LOW (HIGH) RANGE
To proceed at same speed_____	DRIVER STEADY

c. Commands for Control of Turret.

To traverse turret_____	GUNNER TRAVERSE RIGHT (LEFT)
To stop turret traverse_____	STEADY—ON

d. Fire Commands. (See chapter 4.)

Section III. CREW DRILL

53. Dismounted Drill

a. To Form Tank Crew. Being dismounted, the crew takes dismounted posts at the command FALL IN.

b. Fall In. On Command, the crew falls in at attention. The tank commander takes his post two yards in front of the right track, facing the front. The gunner, bog, driver, and loader, in that order, take posts to the left of the tank commander at close interval.

c. To Break Ranks. Being at dismounted posts, the crew breaks ranks at the command FALL OUT. Crew members habitually fall out to the right of the tank.

d. To Call Off. Crew being at dismounted posts, at the command CALL OFF, the members of the crew call off in turn as follows:

Tank commander_____	TANK COMMANDER
Gunner_____	GUNNER
Bow gunner_____	BOG
Driver_____	DRIVER
Loader_____	LOADER

e. To Change Designations and Duties.
  (1) Crew being at dismounted posts, at the command FALL OUT TANK COMMANDER (GUNNER) (DRIVER):
    (*a*) The man designated to fall out moves along the rear of the rank to the left flank position and becomes loader.
    (*b*) The crew members on the left of the vacated post move one position to the right and prepare to call off their new assignments.
    (*c*) The acting tank commander starts calling off as soon as the crew is re-formed in line.
  (2) The movement may be executed by having any member of the crew fall out except the loader.
  (3) All movements should be executed at double time with snap and precision.

54. To Mount Tank Crew

Crew being at dismounted posts:

Note. All phases of crew drill begin with tank gun forward.

Tank commander	Gunner	Bow gunner	Driver	Loader
Command: PREPARE TO MOUNT.				
About face.	About face.	About face.	About face.	About face.
Command: MOUNT.				
Stand fast.	Mount right fender.	Stand fast.	Stand fast.	Mount left fender.
Mount right fender.	Mount right stowage box.	Stand fast.	Stand fast.	Mount left stowage box.
Mount right stowage box.	Enter turret and take post.	Mount right fender.	Mount left fender.	
Enter turret.	Connect breakaway plugs.	Enter bow gunner's hatch.	Enter driver's hatch.	Enter turret and take post.
Connect breakaway plugs.		Connect breakaway plugs.	Turn on master relay switch.	Turn on radio.
Command: REPORT.	Report GUNNER READY.	Report BOG READY.	Connect breakaway plugs.	Connect breakaway plugs.
			Report DRIVER READY.	Report LOADER READY.

55. To Close and Open Hatches

a. To Close Hatches. Crew being at mounted posts:

Tank commander	Gunner	Bow gunner	Driver	Loader
Command; CLOSE HATCHES.	Release turret lock and insure that neither the turret nor the gun blocks hatches.			
Close hatch.		Close hatch.	Close hatch.	Close hatch.
		Raise periscope.	Raise periscope.	Raise periscope.
Command: REPORT.	Report GUNNER READY.	Report BOG READY.	Report DRIVER READY.	Report LOADER READY.

b. *To Open Hatches.* Crew being at mounted posts:

Tank commander	Gunner	Bow gunner	Driver	Loader
Command: OPEN HATCHES.	Insure that neither the turret nor the gun blocks hatches.	Lower periscope.	Lower periscope.	Lower periscope.
Open hatch.		Open hatch.	Open hatch.	Open hatch.
Command: REPORT.	Report GUNNER READY.	Report BOG READY.	REPORT DRIVER READY.	Report LOADER READY.

56. To Dismount Tank Crew

Crew being at mounted posts, hatches open, turret straight forward:

Tank commander	Gunner	Bow gunner	Driver	Loader
Command: PREPARE TO DISMOUNT.				
Disconnect breakaway plugs.	Disconnect breakaway plugs.	Disconnect breakaway plugs.	Disconnect breakaway plugs. Turn off master relay switch.	Disconnect breakaway plugs. Turn off radio.
Command: DISMOUNT.				
Emerge from turret.----	Stand fast.-------------	Emerge from hatch.------	Emerge from hatch.------	Emerge from turret.
Move to right stowage box.	Emerge from turret.------	Move to right fender.-----	Move to left fender.-----	Move to left stowage box.
Move to right fender.---	Move to right stowage box.	Take dismounted post.---	Take dismounted post.---	Move to left fender.
Take dismounted post and command CALL OFF.	Move to right fender. Take dismounted post.	------------------------	------------------------	Take dismounted post.

91

57. To Dismount Through Escape Hatches

Without weapons, crew being at mounted posts:

Tank commander	Gunner	Bow gunner	Driver	Loader
Command: THROUGH ESCAPE HATCHES, PREPARE TO DISMOUNT.				
Disconnect breakaway plugs.	Disconnect breakaway plugs. Traverse turret to give access to forward compartment.	Disconnect breakaway plugs. Open escape hatch.	Disconnect breakaway plugs. Turn on master relay switch. Open escape hatch.	Disconnect breakaway plugs and turn off radio.
Command: DISMOUNT.				
Stand fast.	Move to left side of turret.	Dismount through escape hatch and take dismounted post.	Dismount through escape hatch and take dismounted post.	Stand fast. Enter bow gunner's compartment and dismount.
Move to left side of turret.	Enter bow gunner's compartment and dismount.			Take dismounted post.
Enter bow gunner's compartment and dismount.	Take dismounted post.			
Take dismounted post.				

58. Pep Drill

To vary the drill routine and to maintain the interest of the crew members, unexpected periods of pep drill are introduced into the training. Pep drill consists of a series of precision movements executed at high speed and terminating at the position of attention, either mounted or dismounted. For example, the crews being dismounted, the platoon commander commands: IN FRONT OF YOUR TANKS, FALL IN; MOUNT; DISMOUNT; FALL OUT TANK COMMANDER; ON THE LEFT OF YOUR TANKS, FALL IN; FORWARD, MARCH; BY THE RIGHT FLANK, MARCH; TO THE REAR, MARCH; MOUNT. Preparatory commands for mounting and dismounting are eliminated in this drill. Posts of all crew members are changed frequently.

Section IV. SERVICE OF THE PIECE

59. General

a. The gun crew in the tank consists of the loader, who loads the tank gun and the coaxial machinegun; the gunner, who ranges on the target, aims the guns, and fires; and the tank commander, who controls the fire and, when necessary, adjusts the fire.

b. Teamwork, coordination, and precision of movement are of utmost importance in service of the piece. Crew cooperation in training will provide a smooth, efficient operation in combat when speed is essential and delays or mistakes may be fatal.

60. Gun Crew Positions, Mounted

Tank commander	Right rear of turret.
Gunner	Right side of tank gun.
Loader	Left side of tank gun.

61. Safety Precautions

a. Safety precautions and proper operating procedures are absolutely necessary if the tank is to be kept in operation. The procedures and precautions listed below should be repeated over and over so the normal procedure will be a *safe* procedure.

b. The loader—

　(1) Must check the tank gun continually to insure proper functioning.

　(2) Will check the bore of the gun for obstructions prior to and during firing.

　(3) In loading the gun, must not allow the fuzed nose or the primer of the round to strike any solid object in the turret.

　(4) Will carefully examine each round of tank gun ammunition to see that it is clean and not bulged or dented.

　(5) Will not attempt to remove ammunition from below the turret floor until the loader's traverse safety is off.

(6) Will not attempt to disassemble any portion of a round of tank ammunition unless ordered to install the concrete-piercing fuze on HE ammunition.

(7) Will stay clear of the path of recoil after loading the gun.

(8) Will not attempt to trip the extractors with his fingers when closing the breech.

(9) Will not remove the coaxial machinegun from the tank until it has been cleared and inspected by the tank commander.

(10) Will turn radio off before tank engine is started.

c. The gunner—

(1) Will not fire the main armament or the coaxial machinegun without warning the crew.

(2) Will not traverse the turret in power without alerting the crew.

(3) Will release the hand firing lever, after firing the gun manually, to avoid injury to the loader or damage to the gun as the next round is loaded.

(4) In the event of a misfire, will turn off the 90-mm gun switch and announce MISFIRE.

(5) Will not turn range knob on range finder when commander is laying gun for elevation.

d. The tank commander must know and enforce all necessary safety precautions within his tank.

e. Any individual who observes a condition making firing unsafe will immediately call or signal the command CEASE FIRING.

62. To Open Breech

Grasp the grip portion of the operating handle, release the latch on the grip, and pull the handle to the rear and down. *When the breech is locked open, immediately return the operating handle to its upward position and latch it.*

63. To Load Gun

a. Open the breech and return the operating handle to its latched position.

b. Secure a round of ammunition; grasp it by the base of the shell case with the right hand and in the rear of the ogive with the left hand.

c. Place the projectile in the breech recess, taking care not to strike the fuze. Move the round forward until the projectile rests in the chamber; remove the left hand, and push the round until the projectile is well into the chamber. Then, with the heel of the right hand, vigorously push the round forward into the chamber, rotating the body to the left and sliding the hand off the round upward and to the left to insure clearing the breech. The automatic breechblock will push the hand clear if it should follow the round too far into the breech recess.

Move to the left side of the turret, clear of the path of recoil. Push in on the loader's reset safety (on tanks so equipped) and announce UP.

64. To Unload an Unfired Round or a Misfire

a. To unload an unfired round, the loader cups his hands behind the breech to catch the base of the round as it emerges and to prevent it from dropping to the floor. The gunner, by means of the operating handle, opens the breech *slowly.* (*The breech must not be opened rapidly, or the case will become separated from the projectile.*) The loader then removes the round and returns it to its rack.

b. To unload a misfire, the following precautions will be taken: two more attempts, first electrically and then manually, will be made to fire the piece. Wait 1 minute from the time of the last attempt before opening the breech. Before removing the round, personnel unnecessary to the operation should be cleared from the vicinity. Remove the round. Rounds which misfire will not be returned to the racks, but will be removed to a safe place and turned over to ordnance personnel.

65. To Remove a Stuck Projectile

If the case and projectile become separated despite care in opening the breech, the chamber will be filled with rags to form a cushion so that the projectile will not damage the breechblock. The breech will be closed and the procedure described in paragraph 66 should be followed. After the projectile is free in the chamber, the breech will be opened and the projectile removed and disposed of in accordance with existing regulations. The chamber must be cleaned.

66. To Remove a Stuck Round

When a round is stuck in the gun and it is impossible or inadvisable to fire it out, it will be removed *under the direct supervision of an officer.* With the breech open, the loader takes position to receive the round as it is pushed from the chamber. The bow gunner dismounts, inserts the bell rammer into the muzzle of the gun, and pushes it gently down the bore until it is seated on the ogive of the projectile. Exerting a steady pressure, he shoves the round clear so that it may be removed by the loader. To the maximum possible extent, all parts of the body should be kept clear of the muzzle or breech during the operation. If this procedure fails to remove the round, experienced ordnance personnel should be called. Sometimes the round can be pried out by using the extracting and ramming tool or the base of an empty shell case as a lever.

Section V. MOUNTED ACTION

67. General

Prior to mounted action drill, the following conditions must be met.

a. Crew mounted.

b. Hatches open.

c. Tank gun forward.

d. Turret-mounted machinegun uncovered.

e. Ammunition stowed.

68. Prepare to Fire

A series of checks of turret components must be systematically performed by the tank crew to insure that the equipment is in proper working condition. These checks are performed before every operation. During training, the tank crew is drilled to perform these duties to insure coordination of effort, completeness of checks, and speed of execution. All checks listed must be performed in either the assembly area or attack position. A final check is made just prior to crossing the line of departure. Commands and duties of crewmen are listed below.

Tank commander	Gunner	Bow gunner	Driver	Loader
Command: PREPARE TO FIRE.				
Clean gunner's and tank commander's periscope, end windows of M12 range finder, and vision blocks.	Check recoil oil by physically feeling indicator tape. Clean and inspect M20 periscope. Check instrument light and install batteries.	Clean periscope, lower seat, close hatch.	Clean periscope, lower seat, close hatch.	Open breech and inspect tube and chamber for obstruction and cleanliness. Close breech. Check and adjust headspace on coaxial machine gun. Inspect all ammunition in turret for completeness of stowage and serviceability.
Command: CHECK FIRING SWITCHES.	Turn 90-mm gun switch to ON position.	Check headspace of caliber .30 machinegun.	Turn on master relay switch. Start auxiliary engine.	Turn manual safety to OFF position. Watch action of solenoid and listen for click of percussion mechanism.
Check firing trigger on power control handle. Recock main gun after each firing check.	Check firing trigger on power control handle. Check firing trigger on manual elevation control handle. Check manual firing control.			
	Turn off 90-mm gun switch and turn on coaxial machinegun switch.			Close cover and cock coaxial machinegun. Watch action of solenoid and listen for strike of firing pin.

Tank commander	Gunner	Bow gunner	Driver	Loader
Check firing trigger on power control handle.	Check firing trigger on power control handle. Check firing trigger on manual elevation control. Turn off coaxial machinegun switch.	- - - - -	- - - - -	Recock coaxial machinegun after each check.
Command: CHECK POWER CONTROL.				
	Unlock turret. Check manual traverse (to insure free movement of turret).	- - - - -	- - - - -	Check for obstruction to traverse. Check oil in power system.
	Check manual elevation. Turn on turret motor switch.			
Check power control handle for power elevation and power traverse.	Check power control handle for power elevation and power traverse.			
Check and adjust headspace and timing on turret-mounted caliber .50 machinegun.	Check azimuth indicator for accuracy; traverse turret a complete rotation; coordinate with crew members.	- - - - -	- - - - -	Inspect ammunition in hull stowage. (Coordinate with gunner.)

98

Check sight adjustment; boresight if necessary. Half-load turret-mounted caliber .50 machinegun.	Check sight adjustment; boresight if necessary. Check elevation quadrant by use of M1 gunner's quadrant. Set unit battle sight on M12 range finder and ballistic unit. Await command REPORT.	Place turret in power and check azimuth indicator for slippage. Turn off turret motor switch.	Half-load coaxial machinegun. Open breech. Await command REPORT.
Command: REPORT	Report GUNNER READY.	Report BOG READY.	Report DRIVER READY.	Report LOADER READY.

69. Duties in Firing or Gun Drill

A tank crew must be drilled in the performance of their firing duties to insure coordination of effort and speed of execution. Gun drill is conducted in the form of nonfiring exercises against both stationary and moving targets. Speed must be emphasized throughout this phase of drill. Periods of gun drill must be scheduled and conducted continuously in order to maintain a high standard of tank crew proficiency. To stimulate interest, when possible, the tank should move a few yards between each nonfiring exercise, preferably over a simulated combat course in which there are various types of targets that become visible as the tank advances along the course. For the moving target phase of gun drill, a target mounted on a $\frac{1}{4}$-ton truck can be used. The speed and direction of travel of the target or target vehicle should be varied. The general firing duties of the crew are listed below. For specific firing duties in response to fire commands, see paragraphs 89-109.

Tank commander	Gunner	Bow gunner	Driver	Loader
Be continually alert for targets. Control operation of tank by interphone.	Observe in assigned sector.	Observe in assigned sector.	Observe terrain for best routes. Avoid unnecessary obstacles when possible. Be alert for commands from tank commander.	Observe in assigned sector.
Give fire commands and lay tank gun for direction.				
	Range on target. Lay tank gun for deflection and elevation.	Fire on designated targets.	Observe in assigned sector.	Load ammunition as announced in fire command. Push reset safety on tanks so equipped, and announce UP.
	Fire on target. Adjust fire for target destruction.			
Observe fire and give subsequent fire commands if necessary.				Reload tank gun until CEASE FIRE is given.
If gunner is unable to see target, adjust fire.	Call MISFIRE if tank gun fails to fire.			Follow misfire procedure.

Call STOPPAGE if co-axial machinegun fails to fire.	Reduce stoppage. Fire co-axial machinegun manually if so directed by gunner.
Fire turret-mounted machinegun as necessary.	Keep ammunition available in turret. Refill ready racks as necessary.
	Keep record of ammunition fired.

70. To Clear and Secure Guns

The clear and secure guns procedure, like other procedures in tank operations, is conducted as a drill during training to insure that each crewman knows the duties he must perform in the clearing of the tank weapons and preparing the tank for an administrative move. If it is desired only to clear the tank weapons, the command is CLEAR GUNS. When the weapons are already cleared and it is desired to secure them, the command is SECURE GUNS. When it is desired to perform both phases together, the command is CLEAR AND SECURE GUNS. The crewmen's duties are listed below:

Tank commander	Gunner	Bow gunner	Driver	Loader
Command: CLEAR AND SECURE GUNS.				
Clear turret-mounted machine gun; insert T-block. Inspect periscope.	Turn off gun switches and turret motor switch. Center and elevate tank gun. Inspect periscope and range finder; turn range finder off.	Clear bow gun; insert T-block.	Turn off auxiliary engine--	Clear coaxial machinegun; insert T-block. Clear tank tun; inspect tube and close breech.
Place cover on turret-mounted machine-gun; securet turret-mounted machine-gun in travel lock.	Coordinate with loader to place in travel lock. Place cover on azimuth indicator. Assist loader in placing breech cover on main gun; turn off instrument lights and remove batteries. Await command REPORT.	Place bow gun in travel lock. Place breech cover on bow gun. Open hatch, dismount, and place muzzle cover on main gun and bow gun. Take mounted post and await command REPORT.	Open hatch; raise seat and await command REPORT.	Fill ready racks; open hatch. Secure main gun in travel lock. Place covers over range finder end windows. Assist gunner in placing breech cover on main gun. Resume mounted post; await command REPORT.
Command: REPORT.	Report GUNNER. READY.	Report BOG READY--	Report DRIVER READY.	Report LOADER READY.

71. Loading All Weapons

The tank weapons are loaded on command. This is normally the fire command. The unit SOP may state the type of ammunition to be carried in the chamber of the tank gun before a target appears (par. 29*i*(8)). The machineguns are half-loaded at the command PRE-PARE TO FIRE. In combat, the machineguns will be fully loaded when the unit is deployed for action.

72. Stowage and Handling of Ammunition

a. The ammunition stowage racks in the M47 tank are located under the turret floor and on the right and left sides of the turret in the hull (fig. 39). The ready racks are on the turret floor, and the rounds must

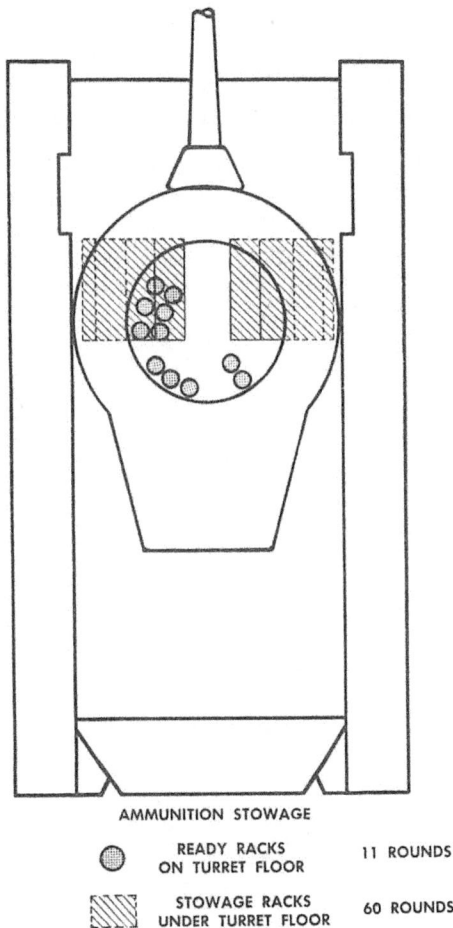

AMMUNITION STOWAGE

⬤ READY RACKS ON TURRET FLOOR   11 ROUNDS

▨ STOWAGE RACKS UNDER TURRET FLOOR   60 ROUNDS

Figure 39. Ammunition stowage.

be removed from the ready racks before the loader has access to the stowage racks. With only 11 rounds immediately available to the loader, the unit SOP should state the number of each type of round to be carried in the ready racks for various combat situations. The remaining 60 rounds should be positioned in the stowage racks so that the most critical type of tank gun ammunition carried in the tank will be readily available to the loader. This also should be prescribed by the unit SOP for various combat situations.

b. Ammunition should be handled in a manner which will prevent striking the projectile or primer of the round against a hard surface. Each round must be inspected for dents or bulges and for dirt before stowing it in the tank. HE ammunition will be received with the fuse set at SQ (superquick) and should so remain until DELAY is required. Because the primer is the most sensitive portion of a round of ammunition, the ammunition should be passed from the ground into the tank with the primer up.

Section VI. DISMOUNTED ACTION

73. To Fight on Foot, Dismounting Through Hatches

a. The crew will be at mounted posts, with tank gun forward, and hatches open. Crew members, including the tank commander, remain below hatches while preparing to dismount and fight on foot. Having completed their preparations, they maintain their positions until the command DISMOUNT is given.

Tank commander	Gunner	Bow gunner	Driver	Loader
Command: PREPARE TO FIGHT ON FOOT. Disconnect breakaway plugs.	Disconnect breakaway plugs.	Disconnect breakaway plugs.	Disconnect breakaway plugs.	Disconnect breakaway plugs.
Order distribution of grenades. Take carbine and six magazines.	------	------	------	Procure grenades.
Take grenades and binocular. Stand fast.	------	Install elevating mechanism on bow gun; dismount gun; install pintle.	Stand fast.	
		Procure grenades; take spare parts roll and spare bolt assembly from gunner.	------	Procure three boxes of caliber .30 ammunition.
	Hand spare parts roll and spare bolt assembly to bow gunner. Stand fast.	------		Procure submachine gun and six magazines of caliber .45 ammunition.

Tank commander	Gunner	Bow gunner	Driver	Loader
Command: DISMOUNT. Dismount via right stowage box and fender.	-------	-------	-------	Pass two boxes of caliber .30 ammunition to tank commander.
Receive two boxes of caliber .30 ammunition from loader.	Take one box of caliber .30 ammunition. Dismount with one box caliber .30 ammunition.	Pass bow gun to driver---	Dismount to right fender and receive bow from bow gunner.	
Act as squad leader of machinegun crew.	Take tripod from right front fender box. Mount tripod. Help mount gun; serve gun as Nr 1.	Dismount with carbine, spare parts roll, and spare bolt assembly. Receive one box caliber .30 ammunition from loader. Serve gun as Nr 3.	Dismount with bow gun. Mount bow gun; serve gun as Nr 2.	Pass one box of caliber .30 ammunition to bow gunner. Dismount and provide security for machinegun crew with submachine gun. (Under certain conditions the loader may remain in the turret, connect the breakaway plugs, and maintain contact with the platoon leader.)

b. The dismounted crew moves to the position indicated by the tank commander. This position during drill is usually 5 yards in front of the tank. The crew members take the posts and perform the duties prescribed for machine-gun drill.

c. In combat, tank personnel will not ordinarily fight on foot. Only in emergencies when it is necessary to defend a disabled tank or evacuate a destroyed tank should tank crewmen fight outside their vehicle. However, since tank personnel are often employed to furnish close-in security for their own vehicles, they must be thoroughly familiar with the use of weapons which can be dismounted from the tank.

74. To Remount After Dismounted Action

Tank commander	Gunner	Bow gunner	Driver	Loader
Command: OUT OF ACTION.				
Supervise taking gun out of action.	Help dismount gun. Fold tripod.	Take mounted post. (Leave caliber .30 ammunition in front of tank.) Receive and mount bow gun. (Remove and stow ground accessories.)	Dismount machine gun. Pass bow gun to bow gunner. Stow tripod.	Resume mounted post. Stow sub-machinegun and ammunition.

Tank commander	Gunner	Bow gunner	Driver	Loader
Pass remaining caliber .30 ammunition to loader.	Mount tank with remainder of caliber .30 ammunition and stow it. Stow spare parts roll and spare bolt assembly.	Hand gunner spare parts roll and spare bolt assembly.	Take mounted post.	Receive remaining ammunition and stow it.
Mount to turret; stow grenades. Stow carbine and ammunition.				Stow grenades.
Connect breakaway plugs.	Connect breakaway plugs.	Connect breakaway plugs.	Connect breakaway plugs.	Connect breakaway plugs.
Command: REPORT.	Report GUNNER READY.	Report BOG READY.	Report DRIVER READY.	Report LOADER READY.

75. To Abandon Tank

If it becomes necessary to abandon tank, the crew proceeds as follows:

a. If time permits deliberate action, the tank commander displays the flag signal DISREGARD MY MOVEMENTS (par. 7, FM 21–60) and supervises the disabling of those weapons remaining in the tank. Back plates are removed from machineguns, and the percussion mechanism is removed from the tank gun. Like items in the spare parts kit are also removed. Individual weapons and maximum amunition loads are carried.

b. Ordinarily, the tank is abandoned as a result of a direct hit which causes it to catch fire or disables it so that it becomes a vulnerable target. At the command ABANDON TANK, crew members open the hatches, dismount, and take cover at a safe distance from the tank. In case of fire, it is particularly important to hold the breath until clear of the vehicle; inhaling fumes and smoke may cause serious injury.

76. To Destroy Tank

When the command DESTROY TANK is given, crew members first remove equipment that is to be carried. They then destroy the tank and the remaining weapons, ammunition, and equipment as prescribed by unit SOP.

Section VII. EVACUATION OF WOUNDED FROM TANKS

77. General

Wounded members of the tank crew normally will evacuate themselves from disabled tanks or will be removed by their fellow crew members. The utmost speed is necessary in order to save the lives of those who are unhurt as well as the life of the casualty. A burning tank can trap its crew in a matter of seconds, and an enemy who has determined the range and has disabled a tank will probably continue shooting until the vehicle burns. It is essential, therefore, that all crew members become extremely proficient in utilizing the quickest methods of removing each other from the tank. Speed is the primary requisite; care in handling will be stressed only where it has been possible to move the tank to cover. If the action has ceased momentarily, or if the tank has been able to disengage itself without hindering the accomplishment of the mission, the casualty is removed on the spot and carried to a protected place where emergency first aid is administered. Otherwise, the action will be continued until an opportunity to remove casualties is presented.

78. Methods of Evacuation

Methods of evacuation described herein are based on the employment of a two-man team, the largest team that can work effectively around a single hatch opening. In some cases, a third man will be able to give considerable help from inside by placing belts around the wounded man or by moving him to a position where he can be grasped from above. Necessity for swift action usually will require that the casualty be removed by grasping his clothing or his arms. If an arm is broken, or if there are other injuries which will be aggravated by these procedures, and if time allows, some form of sling may be improvised which will protect the part from further injury. Any equipment which is immediately available, such as pistol belts, web belts, or field bag straps, will be used for this purpose.

79. Evacuation Drill, General

a. This paragraph contains general information which may be used as a guide in practicing the evacuation of crew members from any position. During drill, the composition of the evacuating team should be changed frequently to provide practice for all members of the crew in meeting various emergencies.

b. The first member of the crew to discover that another is badly wounded calls BOG (LOADER, etc.) WOUNDED. If the tank is not actively engaged and the tank commander decides that evacuation is necessary, he commands EVACUATE BOG (LOADER, etc.). Crew members dismount, unless one man is needed to help from inside, and the two nearest the hatch above the wounded man take stations at that hatch to act as the evacuation crew. If the man nearest the casualty sees that his help from inside the tank is needed, he stays inside and arranges a sling immediately or takes whatever steps he can to speed the operation. Before leaving the wounded man, first aid is administered. The wounded man is moved to a sheltered position, which is marked so that he will not be run over; the tank commander reports by radio that he has lost one or more men and gives the location where they may be found.

Section VIII. INSPECTIONS AND MAINTENANCE

80. General

The tank commander is responsible for insuring that required inspections are made. Mechanical efficiency is essential to tank unit operation; therefore each tank must be inspected systematically at intervals during each day of use. Defects can then be discovered and corrected before they result in mechanical damage or failure. Crew members make their individual inspections and report the results to the tank commander, who, in his own report, lists all items requiring the services of maintenance personnel. In supervising driver maintenance or other services performed at periodic intervals and from day to day, the tank commander delegates responsibility to crew members as necessary. Maintenance procedures omitted from this manual are set forth in detail in TM 9–718A.

81. Maintenance To Be Performed

a. Inspections are made of all personal equipment and weapons, fire-control equipment, communication equipment, vehicle equipment, and mechanical features of the vehicle. Inspections of instruments, lights, tracks, suspension system, and engine performance are made in accordance with provisions of appropriate technical manuals. The driver fills in DD Form 110 (Vehicle and Equipment Operational Record), indicating thereon deficiencies and maintenance work required. He must prepare this form carefully. Any irregularity entered on DD Form 110 and not repaired before the tank is again used will be reentered on the same form continually until the deficiency has been corrected.

b. In succeeding paragraphs (82–86), the duties of crew members in performing tank inspections are tabulated in chart form to be used as an aid in training. However, the steps listed on these charts need not be followed exactly—though they should be used as a guide—in the later conduct of crew drill.

82. Before-Operation Inspection

The tank is locked and covered by a tarpaulin, with gun in traveling position. For training purposes and safety precautions, this inspection and maintenance is divided into three phases. Each phase is completed before beginning the next. The turret is traversed as necessary to facilitate the various operations.

Tank commander	Gunner	Bow gunner	Driver	Loader
Command: FALL IN; PREPARE FOR INSPECTION.	Fall in	Fall in	Fall in	Fall in.
Inspect crew	Stand inspection	Stand inspection	Stand inspection	Stand inspection.
Command: PERFORM BEFORE-OPERATION INSPECTION, PHASE A. Supervise inspection and filling out of DD Form 110 and DA Form 11–238. Inspect tracks and suspension system.			Fill out DD Form 110 during inspection.	Fill out DA Form 11–238 during inspection.
	Help remove, fold, and lay tarpaulin to left side of tank.	Inspect ground beneath tank for fuel or oil leaks.	Help remove, fold, and lay tarpaulin to left side of tank.	Help remove, fold, and lay tarpaulin to left side of tank.
	Mount left front fender; open loader's hatch and enter tank.		Mount left front fender; check contents of outside stowage boxes.	Mount right front fender; check contents of outside stowage boxes; check pioneer tools and tow cable.

Gunner	Bog	Driver	Loader
Unlock commander's hatch; remove breech cover, inspect breech of 90-mm gun.	Mount to rear deck via right front fender; check fuel in filler necks.	Remove tool bag with hand tools; lay tools out on tarpaulin and check for presence and condition.	Remove and stow muzzle cover; unlock gun traveling lock; check muffler clamps; check for turret fouling and for waste or rags around mufflers. Help open power package grilles; check batteries. Assist in checking oil levels.
Move into bow gunner's compartment; unlock and open hatch; remove gun cover and clear bow gun.	Open power package grilles; check batteries. Check oil levels of main engine, auxiliary engine, and transmission; check auxiliary engine air cleaner.	Take mounted post------	Receive and stow breech covers; mount antennas. Take mounted post. Check to see if radios are off.
Unlock driver's hatch. Move into turret; pass all covers to loader; pass antenna sections to loader; unlock turret lock; take mounted post.			
Report GUNNER READY, PHASE A COMPLETED.	Report BOG READY, PHASE A COMPLETED.	Report DRIVER READY, PHASE A COMPLETED.	Report LOADER READY, PHASE A COMPLETED.

Command: REPORT----

Tank commander	Gunner	Bow gunner	Driver	Loader
Command: PERFORM PHASE B. Procure cleaning rods; swab bores of machineguns and turret guns.	Traverse turret manually to check azimuth indicator.	Visually check main engine for security of components.	Check fixed fire extinguishers; turn on master relay switch; set parking brake.	Check portable fire extinguisher; open fuel valves.
Assist with sight adjustment.	Adjust sights; check oil in power traverse reservoir. Check operation of power traverse and elevation and firing controls.	Listen for operation of fuel cutoff; observe operation of auxiliary engine.	Check operation of: heater and blower; manual steering control; service, IR, and blackout lights; horn; bilge pumps; drain valves; and fuel cutoff. Start auxiliary engine. Stop engine after checking.	Check: water, first-aid kit, rations; 90-mm ammunition; armament tools and spare parts; spare recoil and engine oil; air cleaners.
Help driver check lights..	Assist with checking air cleaners.			
Command: REPORT____	Report G U N N E R READY, PHASE B COMPLETED.	Report BOG READY, PHASE B COMPLETED.	Report D R I V E R READY, PHASE B COMPLETED.	Report L O A D E R READY, PHASE B COMPLETED.

114

Command / Crew task			
Command: PERFORM PHASE C. Mount turret machinegun, adjust headspace, check timing, check mount.	Help attach empty cartridge bag. Check sighting and fire-control equipment, main gun, mount, and recoil mechanism. Clean chamber and breech of main gun.	Observe main engine exhaust; visually check engine for oil leaks, vibration, or unusual noises.	Attach empty cartridge bag; assist tank commander in mounting turret machinegun; adjust coaxial machinegun headspace, check mount and firing mechanism; check periscope; turn on radio set. Check hand grenades and caliber .45 ammunition.
		Check transmission oil with main engine at proper idling speed.	
		Close power package grilles.	
			Connect breakaway plug.
Direct driver to move vehicle slowly forward two tank lengths; inspect right track and suspension.		Start main engine; during warmup, check instruments and smoothness of operation; listen for unusual noises. Turn on external interphone equipment. After engine warmup, check magneto operation; idle engine while bog is checking transmission oil.	
	Dismount to left side of tank; inspect track and suspension system while vehicle is moving slowly forward.	Drive vehicle slowly forward two tank lengths.	Perform operator daily maintenance on radio equipment as prescribed in DA Form 11–238.
Direct driver to move vehicle to original position.	Check tools for presence and condition; place in bag and stow.	Drive tank to original position.	
Lock gun traveling lock if gun is in traveling position. Check operation of auxiliary interphone equipment.	Traverse turret to either traveling position or firing position. Connect breakaway plugs.		Check operation of external interphone equipment.

Tank commander	Gunner	Bow gunner	Driver	Loader
Help roll and stow tarpaulin. Take mounted post. Check carbine and ammunition, periscope, vision blocks, power control, binocular, flag set, and operation of radio. Connect breakaway plugs. Check interphone system by commanding: REPORT.		Roll and stow tarpaulin. Take mounted post. Check bow gun, adjust headspace; check escape hatch, periscope, caliber .30 ammunition. Connect breakaway plugs.	Check submachinegun and ammunition, periscope, and escape hatch. Connect breakaway plugs.	
	Report GUNNER READY, PHASE C COMPLETED.	Report BOG READY, PHASE C COMPLETED.	Report DRIVER READY, PHASE C COMPLETED.	Report LOADER READY, PHASE C COMPLETED.
Report READY to platoon leader.				

83. During-Operation Inspection

This is a continuous process for all crew members during operation of the vehicle. All crew members must remain on the alert at all times for unusual noises and conditions, reporting to the tank commander if any are discovered.

Tank commander	Gunner	Bow gunner	Driver	Loader
Check operation of radio and interphone system; observe security of antenna, turret-mounted machinegun, and other visible outside equipment.	Check operation of sighting and fire-control equipment. Also check elevating and traversing mechanism if gun is in firing position.	Observe instruments; check security of bow gun and fixed fire extinguishers.	Observe instruments; check operation of controls.	Check security of equipment in turret; check coaxial machinegun, radio, and portable fire extinguisher.

84. At-the-Halt Inspection.

The length of time for the halt is the basis for determining how much of the following inspection will be completed; priority should be given to items accordingly. The tank commander will be informed of the time allotted to the halt and will indicate to his crew how much time will be allotted for maintenance and inspection and how much for relief of the crew members. As the result of training and experience, the crew will learn what can be accomplished in a given length of time.

117

Tank commander	Gunner	Bow gunner	Driver	Loader
Command: PERFORM AT-HALT INSPECTION. (Use interphone system when applicable.)				
Disconnect breakaway plugs, supervise inspection.	Disconnect breakaway plugs. Release turret lock; check operation of manual traverse, power traverse, and elevation.	Disconnect breakaway plugs. Dismount to rear deck; unlock gun traveling lock (if applicable).	Disconnect breakaway plugs. Fill out DD Form 110 (during-operation and at-halt sections); idle main engine properly; check instruments.	Man turret-mounted gun (if applicable); if not applicable, check, applicable items listed under "loader" in before-operation section.
Dismount to rear deck, help bog open power package grilles.	Check sight adjustment, sighting and fire-control equipment, coaxial gun and mount, firing mechanisms, security of air cleaners.	Open power package grilles; check transmission oil while engine is idling; inspect operation of main engine and security of components.		
Clean outside of all turret periscopes, vision blocks, and range finder end box windows.			Stop main engine	Place radio speaker into operation.

118

Command	Gunner	Bog	Driver	Loader
	Traverse turret to enable bog to lock gun traveling lock.	Inspect operation of auxiliary engine. Check main engine oil level. Close power package grilles.	Start auxiliary engine. Stop auxiliary engine. Check driver's compartment for fuel and oil leaks. Check fixed fire extinguishers, brakes, and controls.	Disconnect breakaway plugs.
	Check security of equipment stowed in turret.	Lock gun traveling lock (if applicable); check pioneer tools and tow cable.	Check service and black-out lights; clean periscope.	
Dismount; inspect track and suspension; check underneath vehicle for fuel or oil leaks; help driver check lights.	Take mounted post	Take mounted post; check bog compartment for fuel and oil leaks; clean periscopes.		Take mounted post when commander is mounted.
Take mounted post				
Command: REPORT.	Report GUNNER READY.	Report BOG READY.	Report DRIVER READY.	Report LOADER READY.

Tank commander	Gunner	Bow gunner	Driver	Loader
Command: FALL OUT FOR BREAK. (If applicable, alternate manning of turret-mounted gun.) Command:				
MOUNT UP (if dismounted). Connect breakaway plugs.	Take mounted post; connect breakaway plugs.	Take mounted post; connect breakaway plugs.	Take mounted post; connect breakaway plugs.	Take mounted post; connect breakaway plugs, turn off radio speaker.
Command: REPORT.	Report GUNNER READY.	Report BOG READY.	Report DRIVER READY.	Report LOADER READY.

85. After-Operation Inspection and Maintenance

Immediately after operation, the tank is given service and maintenance necessary to prepare it in every way for sustained operation. The inspection and maintenance covers all points listed in the before-operation inspection and covers them in practically the same order. Obviously, more extensive servicing and maintenance is required. During this operation, the vehicle is cleaned, serviced, and replenished with fuel, oil, grease, ammunition, first-aid equipment, water, and rations. Refer to the vehicle lubrication order for proper types and amount of oil and greases and intervals of use. All safety precautions against fire must be observed while refueling. A portable fire extinguisher must be available on the rear deck of the tank and manned by a crew member. Also, special safety precautions must be observed in handling ammunition.

Tank commander	Gunner	Bow gunner	Driver	Loader
Command: PERFORM AFTER-OPERATION INSPECTION AND MAINTENANCE. Supervise operation.			Fill out appropriate section of DD Form 110 during inspection.	
Check DD Form 110 and DA Form 11-238 for completeness; make final inspection of vehicle; report READY to platoon leader.	Clean turret gun, bow gun, turret-mounted machine gun, coaxial gun, and inside of turret; replenish ammunition, water, rations, and first-aid equipment.	Help driver clean outside of vehicle; clean engine, engine compartment, and bog compartment; replenish fuel, oil, and greases.	Idle engine properly before stopping; clean outside of vehicle; clean engine, engine compartment, and driver's compartment; replenish fuel, oil, and greases. Complete DD Form 110 and give to tank commander.	Help gunner clean turret gun, bow gun, turret-mounted machine gun, coaxial gun, and inside of turret; replenish ammunition, water, rations, and first-aid equipment. Complete DA Form 11-238 and give to tank commander.

86. Periodic Additional Inspection and Maintenance

This service and maintenance is performed weekly in garrison in addition to after-operation inspection and maintenance. It is also performed after each field operation in combat and on maneuvers. In garrison operation, allowances should be made for these services in preparing training schedules and work details. In combat and on maneuvers, provisions should be made (where possible) to allow time for crew members to perform this inspection and maintenance.

Tank commander	Gunner	Bow gunner	Driver	Loader
Command: FALL IN, PREPARE FOR INSPECTION. Inspect crew.	Fall in.	Fall in.	Fall in.	Fall in.
Supervise inspection.	Stand inspection.	Stand inspection.	Stand inspection.	Stand inspection.
	Clean and paint any rusty spots in turret.	Help driver clean engines and compartments; make detailed inspection of main and auxiliary engines; service batteries.	Clean engines and compartment; make detailed inspection of main and auxiliary engines; service batteries.	
	Tighten all track nuts; inspect track and suspension system.	Procure hand and pioneer tools, clean and check; clean and spot-paint rusty spots on outside of vehicle.	Start main engine and drive vehicle forward as required, to tighten track nuts; clean and paint rusty spots in driver's and bog's compartments.	Help gunner tighten track nuts; inspect track and suspension system.
Check DD Form 110 and DD Form 11-238 for completeness; make final inspection of vehicle; report READY to platoon leader.	Help lubricate.	Help lubricate.	Lubricate as required. Complete DD Form 110 and give to tank commander.	Help lubricate.

Section IX. DESTRUCTION OF EQUIPMENT AND STOWAGE

87. Destruction of Equipment

a. The destruction of materiel is a command decision to be carried out only on authority delegated by the division or higher commander. This usually is made a matter of standing operating procedure. *It is ordered only after every possible measure for preservation or salvage of the materiel has been taken and when, in the judgment of the military commander concerned, such action is necessary to prevent—*

(1) Its capture intact by the enemy.

(2) Its use by the enemy, if captured, against our own or allied troops.

(3) Its abandonment in the combat zone.

(4) Knowledge of its existence, functioning, or exact specifications from reaching enemy intelligence agencies.

b. The principles followed are—

(1) Methods for the destruction of materiel subject to capture or abandonment in the combat zone must be adequate, uniform, and easily followed in the field.

(2) Destruction is as complete as possible within limitations of time, equipment, and personnel. If thorough destruction cannot be completed, the most important features of the materiel are destroyed, and parts which cannot be easily duplicated and are essential to the operation or use of the materiel are ruined or destroyed. *The same essential parts are destroyed on all like units to prevent the enemy's constructing one complete unit from several damaged ones.*

c. Crews are trained in employing prescribed methods of destruction. *Training does not involve actual destruction of materiel.*

d. Certain methods of destruction require special tools and equipment, such as TNT and incendiary grenades, which may not be items of issue. The issue of such special tools and material and the conditions under which destruction will be effected are command decisions and depend upon the tactical situation.

e. The proper methods for the destruction of the OVM and the tank are covered in TM 9–718A.

88. Stowage

a. Importance. Proper stowage of tank equipment is necessary for the efficient functioning of the tank and crew. First, each crew member must ascertain whether or not the equipment necessary to perform his duties is present. Second, and equally important, this equipment must be stowed in the proper place in order to be available when needed. So that these conditions can be met, a list of vehicle stowage and special tools has been prepared for the vehicle. The list of vehicle stowage and special tools designates an exact location either on or within the tank for every piece of authorized equipment, including personal equipment.

b. Stowage List.

Name of part	Quantity required per vehicle	Where carried
AMMUNITION:		
ROUNDS FOR CARBINE, cal. .30, M2 (in 3 30-rd mag).	90	In turret bulge in case.
ROUNDS FOR GUN, Machine, cal. .30 (in metallic link belts).	11, 150	4 250-rd boxes, foot of driver; 4 250-rd boxes on foot of bulkhead; 3 550-rd boxes (bow gun) in bow gunner's compartment; 8 250-rd boxes in turret rack; 1 750-rd box (coaxial gun); 10 250-rd boxes, left rear fender; 5 250-rd boxes, left front fender; 2 250-rd boxes, right front fender; 2 250-rd boxes, right rear fender.
ROUNDS FOR SUBMACHINE-GUN, cal. .45, M3A1 (in 6 30-rd mag).	180	In turret bulge in case.
ROUNDS FOR GUN, Machine, cal. .50 (in metallic link belts).	1700	6 100-rd boxes, rear turret stowage box; 5 100-rd boxes, turret basket floor; 4 100-rd boxes, air cleaner recess; 2 100-rd boxes, hull center tunnel foot.
ROUNDS FOR GUN, 90-mm, M36.	71	60 in center hull, 11 in turret.
GRENADE, Hand_____	(8)	In box, grenade, in turret bulge.
Fragmentation, with fuse_____	4	
Incendiary, with fuse_____	2	
Smoke (WP), with fuse_____	2	
SIGNAL, Ground Pyrotechnic_____	(12)	In box, flare, in turret bulge.
FLARE, White Star, Parachute, M17A1.	2	
FLARE, White Star, Cluster, M18A1.	2	
FLARE, Green Star, Parachute, M19A1.	2	
FLARE, Green Star, Cluster, M20A1.	2	
FLARE, Amber Star, Parachute, M21A1.	2	
FLARE, Amber Star, Cluster, M22A1.	2	

Name of part	Quantity required per vehicle	Where carried
ARMAMENT:		
CARBINE, cal. .30, M2, with sling	1	In turret bulge in bracket.
GUN, Machine, Browning, cal. .30, M1919A4 (flexible).	1	In bow mount.
GUN, Machine, cal. .50, Browning, HB, M2 (turret-mounted).	1	On turret roof.
GUN, Machine, cal. .30 Browning, M1919A4 (coaxial).	1	In coaxial mount.
GUN, Submachine, cal. .45, M3A1, with sling.	1	In turret bulge in bracket.
GUN, 90-mm, M36	1	In turret.
MOUNTS, GUN:		
MOUNT, Machinegun, cal. .50	1	On turret roof.
MOUNT, Tripod, Machinegun, cal. .30, M2.	1	In right rear fender box.
EQUIPMENT FOR GUN, Machine, cal. .30, Browning, M1919A4:		
BRUSH, Cleaning, Chamber, M6	1	In gun spare parts roll.
BRUSH, Cleaning, cal. .30	8	In gun spare parts roll.
CASE, Cleaning Rod, cal. .30, M1	1	In gun spare parts roll.
COVER, spare barrel	2	On spare barrel, hull right sidewall.
EXTRACTOR, Ruptured Cartridge, MK IV.	2	In oddment tray in turret (1—bow gun, 1—coaxial gun).
ROD, Cleaning, Jointed, M1	1	In Case, Cleaning Rod, M1.
WRENCH, Combination, M6	1	In gun spare parts roll.
BAG and adapter, clip, assembly (bow).	1	On bow gun.
BAG, empty cartridge, for gun, Machine, cal. .30 (bow).	1	On bow gun.
BODY, ammunition feedway (bow)	1	On bow gun.
BAG, empty cartridge, assembly	1	On coaxial gun.
CLAMP, muzzle cover	1	On bow gun muzzle cover.
COVER, muzzle, cal. .30, assembly	1	On right front hull (outside).
HOOD, tripod mount, cal. .30 light machinegun.	1	On tripod mount.
EQUIPMENT FOR GUN, Machine, cal. .50, Browning, M2, HB:		
BRUSH, Cleaning, cal. .50, M4	4	In gun spare parts roll.
CASE, Jointed Cleaning Rod and Brush, M15.	1	In gun spare parts roll.
EXTRACTOR, Ruptured Cartridge Case, cal. .50.	1	In oddment tray.
GAGE, headspace and timing, cal. .50.	1	In gun spare parts roll.
HIDER, Flash (M2)	1	On turret-mounted gun.
ROD, Cleaning, Jointed, cal. .50, M7.	1	In Case, Cleaning, M15.
WRENCH, muzzle gland and adjusting screw.	1	In gun spare parts roll.

Name of part	Quantity required per vehicle	Where carried
EQUIPMENT FOR GUN, Machine, cal. .50, Browning, M2, HB—Con.		
COVER, spare barrel, cal. .50_____	1	On left outside of turret bustle.
COVER, machinegun, cal. .50 (turret-mounted).	1	On gun.
EQUIPMENT FOR CARBINE, cal. .30, M2: CASE, ammunition.	1	In turret bulge, in bracket.
EQUIPMENT FOR GUN, Submachine, cal. .45, M3A1:		
CASE, ammunition_____	1	In turret bulge, in bracket.
EQUIPMENT FOR GUN, 90-mm, M36:		
ADAPTER, bore brush and wiper ring.	1	In rear turret stowage box.
BRUSH, Bore, 90-mm, M19_____	2	In rear turret stowage box.
COVER, Canvas, Brush, Bore, M518.	2	On brush, bore.
COVER, Gun Book, M539_____	1	In pamphlet bag.
EYE, lifting_____	1	In gun spare parts roll.
FORM, Artillery Gun Book or gun record book.	1	In cover, gun book, under commander's footrest.
HEAD, Rammer, Unloading, M16__	1	In rear turret stowage box.
OIL, Hydraulic, 1 quart can_____	1	In right rear fender box.
RING, wiper_____	1	In rear turret stowage box.
ROD, push, assembling and disassembling, shaft diameter ½ in, length 6 in. pt ⅛ in.	1	In gun spare parts roll.
ROPE, manila, 3-strand, ½ in. diameter, 6 ft long (for lowering breechblock).	1	In oddment tray.
STAFF SECTION T 3, (alum)_____	5	In right rear fender box.
TOOL, breechblock removing_____	1	In gun spare parts roll.
WRENCH, Fuze, M18_____	1	In rear turret stowage box.
WRENCH, spanner, face, pin type, c to c of pins 2 in., diameter of pin ¼ in., length 6¾ in.	1	Gun spare parts roll.
WRENCH, tubular, single-end, 2 prongs, with pin handle, OD, 1¾₂ in.	1	In gun spare parts roll.
COVER, breech, assembly_____	1	On gun.
COVER, muzzle, assembly_____	1	On gun.
GUN (filler), oil (recoil), hand-operated, suction type (complete with hose assembly), capacity 2 oz.	1	In gun spare parts roll.
SPARE PARTS FOR GUN, Machine, cal. .30 Browning, M1919A4:		In box in gun spare parts roll.
BOLT, assembly_____	1	
BOX, spare parts (empty)_____	1	
EXTRACTOR, assembly_____	1	
LEVER, cocking_____	1	

Name of part	Quantity required per vehicle	Where carried
SPARE PARTS FOR GUN, Machine, cal. .30 Browning, M1919A4—Con.		
PIN, cocking lever_____	1	
PIN, firing, assembly_____	1	
ROD, driving spring, assembly_____	1	
SEAR_____	1	
SPRING, driving_____	1	
SPRING, sear, assembly_____	1	
TRIGGER_____	1	
BARREL, assembly_____	2	In hull right side of cover.
SPARE PARTS FOR GUN, Machine, cal. .50, Browning, M2, HB:		In box in gun spare parts roll.
BOLT, alternate feed, assembly____	1	
BOX, spare parts (empty)_____	1	
EXTENSION, firing pin, assembly_	1	
EXTRACTOR, assembly_____	1	
LEVER, cocking_____	1	
LOCK, accelerator stop_____	1	
PIN, cocking lever, assembly_____	1	
PIN, firing_____	1	
SEAR_____	1	
SLIDE, sear_____	1	
SPRING, sear_____	1	
STOP, accelerator_____	1	
SWITCH, bolt_____	1	
BARREL, assembly_____	1	In cover outside of turret.
SPARE PARTS FOR MOUNT, Gun, Combination, 90-mm:		
GASKET, copper, soft (recoil piston bearing sleeve plug).	1	In gun spare parts roll.
PLUG (recoil mechanism filler)_____	2	In gun spare parts roll.
PLUG, pipe, sq-hd, ⅛ in. (black)____	1	In gun spare parts roll.
PLUG, recoil cylinder replenisher___	1	In gun spare parts roll.
SPARE PARTS FOR GUN, 90-mm, M36:		
MECHANISM, percussion, assembly.	1	In gun spare parts roll.
BINOCULAR, M17A1, with Case, Carrying, M24.	1	On right turret wall.
QUADRANT, Gunner's, M1, with Case, Carrying, M82.	1	On right turret wall.
LIGHT, Instrument, M30 (for ballistic drive and elevation quadrant).	1	In turret, in clips.
LIGHT, Instrument, M36 (for Periscope, M20).	2	In turret, in clips.
PERISCOPE, M13 (loader's, driver's, and assistant driver's).	5	1 in turret; 2 in hull; 2 spare in driver's compartment.
*PERISCOPE, M19*_____	1	In driver's compartment.
*PERISCOPE, M20*_____	2	In turret.
QUADRANT, Elevation, M13 (on Ballistic Drive, M3).	1	In turret.

Name of part	Quantity required per vehicle	Where carried
Setter, Fuse, M27 (wrench)_____	1	In gun spare parts roll.
TABLE, Firing, FT–90–2 (Abridged), 1950.	1	In pamphlet bag.
TRANSMITTER, Superelevation, M23__	1	In turret (on top of oil gear).
SPARES:		
HEAD, Periscope, M19, assembly__	1	In driver's compartment.
HEAD, Periscope, M20, assembly__	2	In turret.
LAMP, Electric, 3 v, No. 323_____	6	In box.
For Ballistic Drive, M3_____	4	
For Light, Instrument, T22_____	2	
LAMP, Electric, 3 v, No. 325 (for Instrument Light, M36).	4	Packaged separately in box.
LAMP, Electric, 24–28 v, No. 313 (for gun ready light).	4	Packaged separately in box.
LAMP, Electric, 24–28 v, 21 cp, special (for Range Finder, M12).	8	In spare lamp box on range finder.
EQUIPMENT, MISCELLANEOUS:		
BATTERY, Flashlight, BA–30_____	14	
To be put in flashlights_____	8	
To be put in Instrument Lights, M36 and M30.	6	
CANTEEN, M1910, complete with cup and cover.	5	Near crew position. 2 in hull; 3 in turret.
CAN, water, 5 gal, std _____	2	1 left-hand rear outside bustle box; 1 right-hand rear outside bustle box.
CARRIER, Wire Cutter, M1938___	1	In tool bag.
CUTTER, Wire, M1938_____	1	In tool bag in Carrier, Wire Cutter, M1938.
FLAG SET, M238_____	1	In left rear fender box.
CASE, CS–90_____	1	
FLAG, MC–273 (red)_____	1	
FLAG, MC–274 (orange)_____	1	
FLAG, MC–275 (green)_____	1	
FLAGSTAFF, MC–270_____	3	
RATIONS, field type, 3-day supply for 5 men.	15	5 in turret bustle, 2 under turret platform rear of tunnel, and 8 in left rear fender box.
ROLL, bedding, for each man_____	5	
BLANKET, Wool, OD, M1934__	2	In cargo packs.
PIN, tent, shelter, wood_____	5	In cargo packs.
POLE, tent, single section_____	1	In cargo packs.
TENT, shelter, half (new type)__	1	In cargo packs.
STOVE, gasoline, 1-burner, with Case, M1942.	2	In rear turret stowage box.
TUBE, flexible nozzle, cam type____	2	In rear turret stowage box.
ADAPTER, lubrication, gun, hy-draulic-to-push type, thin stem, with locking sleeve.	1	In tool bag.

Name of part	Quantity required per vehicle	Where carried
EQUIPMENT, MISCELLANEOUS— Continued		
BAG, pamphlet, assembly_____	1	In turret, under commander's footrest.
BOX, flare_____	1	In turret bustle.
CONNECTOR (fixed fire extinguisher.	3	In tool bag.
CORD, light extension, inspection, 15-foot, single-contact socket and plug.	1	In tool bag.
COVER, azimuth indicator, assembly.	1	On indicator.
COVER, slip ring, assembly_____	1	Slip ring guard on basket floor.
EXTENSION, lubrication gun, hydraulic-to-hydraulic, 12 in. long.	1 1	In tool bag. In tool bag.
EXTINGUISHER, fire, carbon dioxide, 5 lb, portable.	1	In driver's compartment.
GUN, lubrication, hand lever operated, high pressure, 15 oz capacity, with 6 in hydraulic extension.	1	Left-hand fender box.
LAMP, electric (inspection) 24–28 v, 32 cp, No. 1683.	1	In inspection cord light.
PADLOCK SET, 1¾ in. with clevis, keyed interchangeably, composed of 4 locks and 6 keys.	1 set	1 on loader's hatch; 1 on assistant driver's hatch; 1 on driver's hatch; 1 on right front fender box.
PAULIN, canvas, 12 ft by 12 ft__	1	On rear stowage box.
PUMP and hose assemblies_____	1	In right rear fender box.
OILER, pump, bent spout, 1 pt capacity.	1	On left side floor.
ROLL, gun spare parts and tools___	1	On right side of turret bustle.
TAPE, adhesive, pressure sensitive, water resistant, OD 7, 4 in., 15-yd roll.	1	In left rear fender box.
TAPE, friction, ¾ in. wide, 8 oz roll.	1	In tool bag.
UNION (fixed fire extinguisher)____	3	In tool bag.
WIRE, soft iron, diameter 0.080 in (10 ft).	1	In tool bag.
PUBLICATIONS:		
FORM (envelope), DA AGO 478__	1	In pamphlet bag.
MANUAL, Technical, 9–718A_____	1	In pamphlet bag.
ORDER, Lubrication, 9–7010_____	1	In pamphlet bag.
DIAGRAM, strap location_____	1	In pamphlet bag.
SPARE PARTS VEHICULAR:		
LAMP, electric, 24–28 v, 3 cp, No. 1251.	2	In box 7021398.
LAMP, electric (for dome and instrument lights), 24–28 v, 6 cp, No. 623.	2	In box 7021398.

Name of part	Quantity required per vehicle	Where carried
SPARE PARTS VEHICULAR—Con.		
LAMP, electric, 3 v, No. 323 (for azimuth indicator).	3	In box 7021398.
SHOE, track assembly_____	2	On turret.
TOOLS AND TOOL EQUIPMENT:		
BAG, tool, empty_____	1	In rear stowage box.
BAR, cross socket wrench, $\frac{7}{16}$ in. diam, 8 in. long.	1	In tool bag.
BAR, socket wrench, extension, $\frac{3}{4}$ in. square-drive, 16 in. long.	1	In tool bag.
BAR, jimmy, $\frac{1}{2}$ in. blade width, $11\frac{7}{8}$ in. long.	1	On right side of turret bulge.
BAR, socket wrench, extension, $\frac{1}{2}$ in. square-drive, 5 in. long.	1	In tool bag.
BAR, socket wrench, extension, $\frac{1}{2}$ in. square-drive, 10 in. long.	1	In tool bag.
BAR, socket wrench, extension, $\frac{3}{4}$ in. square-drive, $4\frac{5}{8}$ in. long.	1	In tool bag.
CABLE, towing, $1\frac{1}{8}$ in. wire rope diameter, 10 ft. long.	1	On rear of hull.
CHISEL, machinist, hand cold, $\frac{3}{4}$ in. cut, 8 in. long.	1	In tool bag.
FILE, AS, hand, smooth cut, 10 in__	1	In tool bag.
FILE, AS, three square, smooth cut, 6 in.	1	In tool bag.
HAMMER, Machinist, ball peen, 2 lb.	1	In tool bag.
HANDLE, socket wrench, T-sliding, $\frac{1}{2}$ in. square-drive, 9 in. long.	1	In tool bag.
HANDLE, socket wrench, T-sliding, $\frac{3}{4}$ in. square-drive, 17 in. long.	1	In tool bag.
HANDLE, socket wrench, hinged, $\frac{1}{2}$ in. square-drive, 18 in. long.	1	In tool bag.
HANDLE, socket wrench, ratchet, reversible, $\frac{1}{2}$ in. square-drive, 15 in. long.	1	In tool bag.
HANDLE, socket wrench, speeder, brace type, $\frac{1}{2}$ in. square-drive, 12 in. long.	1	In tool bag.
JOINT, socket wrench, universal, $\frac{1}{2}$ in. square-drive.	1	In tool bag.
PLIERS, combination, slip joint, with cutter, 8 in. long.	1	In tool bag.
PLIERS, lineman's side cutting, 8 in. long.	1	In tool bag.
SCREWDRIVER, Machinist, 5 in. blade.	1	In tool bag.
SCREWDRIVER, special purpose, $1\frac{1}{2}$ in. blade.	1	In tool bag.
WRENCH, adjustable, single open end, $1\frac{5}{16}$ in. jaw opening, 8 in. long.	1	In tool bag.

Name of part	Quantity required per vehicle	Where carried
TOOLS AND TOOL EQUIPMENT— Continued		
WRENCH, adjustable, single open end, 1 5/16 in. jaw opening, 12 in. long.	1	In tool bag.
WRENCH, engineers, double open end, alloy steel, 5/16 and 3/8 in. openings.	1	In tool bag.
WRENCH, engineers, double open end, alloy steel, 7/16 and 1/2 in. openings.	1	In tool bag.
WRENCH, engineers, double open end, alloy steel, 9/16 and 11/16 in. openings.	1	In tool bag.
WRENCH, engineers, double open end, alloy steel, 5/8 and 3/4 in. openings.	1	In tool bag.
WRENCH, engineers, double open end, alloy steel, 13/16 and 7/8 in. openings.	1	In tool bag.
WRENCH, engineers, double open end, alloy steel, 15/16 and 1 in. openings.	1	In tool bag.
WRENCH, set or cap screw, 3/32 in. hex.	1	In tool bag.
WRENCH, set or cap screw, 1/8 in. hex.	1	In tool bag.
WRENCH, set or cap screw, 3/16 in. hex.	1	In tool bag.
WRENCH, set or cap screw, 1/4 in. hex.	1	In tool bag.
WRENCH, set or cap screw, 5/16 in. hex.	1	In tool bag.
WRENCH, set or cap screw, 3/8 in. hex.	1	In tool bag.
WRENCH, set or cap screw, 5/8 in. hex.	1	In tool bag.
WRENCH, socket, 1/2 in. square-drive, 3/8 in., 8 point (detachable).	1	In tool bag.
WRENCH, socket, 1/2 in. square-drive, 7/16 in., 12 point (detachable)	1	In tool bag.
WRENCH, socket, 1/2 in. square-drive, 1/2 in., 12 point (detachable).	1	In tool bag.
WRENCH, socket, 1/2 in. square-drive, 9/16 in., 12 point (detachable).	1	In tool bag.
WRENCH, socket, 1/2 in. square-drive, 5/8 in., 12 point (detachable).	1	In tool bag.
WRENCH, socket, 1/2 in. square-drive, 3/4 in., 12 point (detachable).	1	In tool bag.
WRENCH, socket, 1/2 in. square-drive, 7/8 in., 12 point (detachable).	1	In tool bag.

Name of part	Quantity required per vehicle	Where carried
TOOLS AND TOOL EQUIPMENT—Continued		
WRENCH, socket, ½ in. square-drive, ¹⁵⁄₁₆ in., 12 point (detachable).	1	In tool bag.
WRENCH, socket, ½ in. square-drive, 1 in., 12 point (detachable).	1	In tool bag.
WRENCH, socket, ½ in. square-drive, 1¹⁄₁₆ in., 12 point (detachable).	1	In tool bag.
WRENCH, socket, ½ in. square-drive, 1⅛ in., 12 point (detachable).	1	In tool bag.
WRENCH, socket, ¾ in. square-drive, 1⁵⁄₁₆ in., 12 point (detachable).	1	In tool bag.
WRENCH, socket, ¾ in. square-drive, 1⁷⁄₁₆ in., 12 point (detachable).	1	In tool bag.
WRENCH, socket, ¾ in. square-drive, 1½ in., 12 point (detachable).	1	In tool bag.
WRENCH, center guide bolt (socket, detachable, ¾ in. square-drive, 12 point, size of opening 1¼ in.).	1	In tool bag.
TOOLS, PIONEER:		
AXE, chopping, single bit, 4 lb_____	1	On right rear fender box in bracket.
Bar, crow, pinch point, 60 in_____	1	On left fender.
HANDLE, mattock, 36 in. long_____	1	On right rear fender box in bracket.
MATTOCK, pick, without handle, 5 lb.	1	On right rear fender box in bracket.
SHOVEL, general purpose, round point, D-handle.	1	On right rear fender box in bracket.
SLEDGE, blacksmith's, double face, 10 lb.	1	On outside right turret wall.
TOOLS SPECIAL:		
FIXTURE, track connecting and link pulling.		In rear turret stowage box.
FIXTURE, right_____	1	
FIXTURE, left-hand_____	1	
WRENCH, engineers, 15-degree angle, single open end, carb-S, size of opening 5 in., length 42 in.	1	In right rear fender box.
WRENCH, track slack adjusting (hook spanner), diameter of circle 7⅜ in., length 13½ in.	1	In tool bag.

CHAPTER 4

CONDUCT OF FIRE

Section I. FIRING DUTIES, FIRE COMMANDS, SENSINGS, AND MOVING TARGETS

89. Introduction

This chapter is devoted to conduct of fire for the Tank, 90-mm Gun, M47. Speed and accuracy are of paramount importance when engaging targets with tank weapons, and the desirability of first-round hits cannot be minimized. A well-trained crew accomplishes its mission of destroying enemy armor, materiel, and equipment by teamwork, coordination, and the ability to engage targets rapidly and accurately. To effect coordination, the tank commander issues fire commands using a standard terminology and sequence which controls the actions of the crew. This chapter is a discussion of the method and procedure to engage targets with tank weapons, to include the initial fire command, sensings, primary method of adjustment, alternate method of adjustment, subsequent fire commands, firing at stationary and moving targets, the employment of tank machineguns, special conditions, and the Tank Gunnery Qualification Course. This chapter also covers the use of the "battle sight." For basic principles of conduct of fire, see FM 17-12.

90. Firing Duties of Crew, M47 Tank

Below is a list of firing duties as they are performed by individual crew members.

Crew Member	Firing duties
Driver	Moves the tank as directed by the tank commander.
Bog	Fires the bow caliber .30 machinegun at targets of opportunity and as the tank commander directs.
Loader	Loads the tank gun and coaxial machinegun. Controls the loader's traverse safety switch, manual safety, and loader's reset safety (on tanks so equipped). Reduces stoppages. Inspects, loads, cleans, and stows ammunition.
Gunner	Aims and fires the tank gun and coaxial machinegun through control and use of the range finder, M20 periscope, ballistic unit, gun switches, traverse and elevation controls, and firing triggers. Adjusts fire of tank gun and coaxial machinegun.

Crew Member	Firing duties
Tank commander_____	Controls the movement of the tank and actions of the crew. Gives initial fire commands and subsequent fire commands when necessary. Lays the gun for direction using the commander's power control handle and M20 periscope. Is responsible for selection of targets and volume of fire. Supervises and adjusts fire.

91. Initial Fire Command, M47 Tank

a. The initial fire command contains the necessary information to load, aim, and fire. The tank commander lays the gun for direction while issuing the initial fire command. The loader will load, and the gunner will range on the target using the range finder. When the range finder is inoperative, the gunner will place on the ballistic unit the range announced by the tank commander. The gunner will take the target under fire and will adjust the fire by observing through the direct-fire sight. An example of an initial fire command, showing sequence of elements and duties of the crew, follows:

Element	Command	Duties of crew
Alert_____	GUNNER_____	All crew members prepare to follow the fire command. The gunner, in this case, is specifically designated as the crewman to engage the target. The gunner utilizes the range finder as the primary direct-fire sight and the M20 periscope as the secondary direct-fire sight. The tank commander starts laying the gun for direction, using his power control handle and M20 periscope.
Type of ammunition (and fuze if necessary).	SHOT_____	The loader selects and loads the type of ammunition announced, clears the path of recoil, pushes the loader's reset safety (on tanks so equipped), and announces UP.

Note. The 90-mm gun switch must be ON for the loader's reset safety to operate.

When using the range finder, the gunner indexes the proper code number on the ammunition scale, turns on the gun switch, and observes the target area through the direct-fire sight. When using the M20 periscope, the gunner mentally notes the ammunition announced, turns on the gun switch, and prepares to set off the range to be announced by the tank commander.

Element	Command	Duties of crew
Range (When using the range finder as a ranging instrument, this element is omitted.)	900_ _ _ _ _ _ _ _ _ _	The gunner, when using the range finder as a direct-fire sight rather than as a ranging instrument, places the announced range on the range scale. When using the M20 periscope, the gunner indexes the announced range on the range drum of the ballistic unit.
Direction (Normally this element will be omitted. It will be given only when the commander is unable to lay for direction.)	TRAVERSE RIGHT. STEADY—ON	When the tank commander is unable to lay the gun for direction, the gunner, upon hearing the directional element, traverses rapidly in the direction announced by the tank commander. The tank commander uses the M20 periscope to lay the gun for direction, giving STEADY when the aiming cross is approaching the target. At this time the gunner reduces the rate of traverse. ON is announced when the gun is laid for direction.
Description_ _ _ _ _ _ _ _ _ _ _ _	TANK_ _ _ _ _ _ _ _	The gunner, upon hearing the target description and observing the target, announces IDENTIFIED, at which time the tank commander releases his power control; the gunner then ranges on the target and makes the final precise lay. When the M20 periscope is used, the gunner, after announcing IDENTIFIED, elevates or depresses the gun to place the aiming cross of the sight reticle on the center of the target.
Leads (Used only when engaging moving targets.)	ONE LEAD_ _ _	After ranging on the target or after placing the announced range on the ballistic unit, the gunner applies the announced lead (one lead is 5 mils), using the center of the target as the aiming point. The gunner maintains this lead by smooth, even tracking before, during, and after firing.
Command to fire_ _ _ _ _ _ _	FIRE_ _ _ _ _ _ _ _	After making the final precise lay, the gunner announces ON THE WAY, pauses one second, and fires.

b. The initial fire command to engage a target using the battle sight contains essentially the same elements as above; however, the duties of the crew have been modified. The battle sight is designed to engage dangerous surprise or rapidly fleeing targets. It is imperative that the first round be fired immediately; therefore the advantage to be gained by ranging on the target is sacrificed for speed. The round in the chamber is fired using the battle sight settings. If desired, ammunition may be changed after the first round has been fired (par. 93c). A dangerous surprise target is any target capable of seriously damaging or destroying your tank and which has fired or is about to fire at you. A fleeing target is one which is about to escape or reach a concealed position. An example of a battle sight fire command, showing sequence of elements and duties of the crew, follows:

Element	Command	Duties of crew
Alert_____	GUNNER_____	Alerts all crew members to follow the fire command. The gunner, in this case, is specifically designated as the crewman to engage the target. The gunner utilizes the range finder as the primary direct-fire sight and the M20 periscope as the secondary direct-fire sight. The tank commander starts laying the gun for direction, using his power control handle and M20 periscope.
Ammunition Range_____	BATTLE SIGHT	This term automatically includes both the range and ammunition element, since the predetermined settings have been placed on the direct-fire sights. The gunner turns on the 90-mm gun switch and immediately starts looking for the target. The loader selects another round of the same type ammunition specified in the battle sight.
Direction_____	DIRECT FRONT	This element will normally be omitted, since the tank commander will lay the gun for direction while issuing the initial fire command. If, however, the gun is laid on or very near the target, the term DIRECT FRONT might be used.
Description_____	TANK_____	The gunner, upon hearing the target description and observing the target, announces IDENTIFIED, at which time the tank commander releases his power control. The gunner then makes the precise lay on the target.
Leads (Used only when engaging moving targets.)	ONE LEAD___	The gunner applies the announced lead (5 mils) using the center of mass of the target as the aiming point.

136

Element	Command	Duties of crew
Command to fire_____	FIRE_____	After making the final precise lay, the gunner announces ON THE WAY and fires without delay.

92. Sensings

Rounds are sensed in relation to the target. The tank commander and gunner will mentally sense each round for range and deflection. These sensings are not announced unless the gunner fails to observe the burst or tracer through his direct-fire sight, in which case the gunner will announce LOST. The tank commander will then announce his range sensing only, and his subsequent fire command. The five possible range sensings are listed below.

a. Target. In tank gunnery a round is sensed as TARGET only when the round is observed to actually strike the target and to cause the target to change shape or move, or to cause pieces to fly off or disappear. When shot strikes a metal object, there is usually a distinctive orange flash.

b. Over. A round is sensed as OVER when the burst appears beyond the target or the tracer passes above the target. This sensing is readily identified when firing HE, since the burst seems to silhouette the target.

c. Short. A round is sensed SHORT when either the burst or point of strike is observed between the gun and the target. The target is sometimes temporarily obscured by smoke and dust.

d. Doubtful. A round is sensed as DOUBTFUL when the burst appears to be correct for range but off in deflection, or when the tracer passes to the right or left of the target but apparently is correct for range. (A range change is not made on a DOUBTFUL sensing.) A deflection correction is normally sufficient to secure a target hit.

e. Lost. A round is sensed as LOST when the tank commander or gunner fails to observe the point of strike, burst, or tracer. The point of strike may not be visible due to obscuration, terrain, or failure of the round to detonate. (Based on his judgment of the terrain, the tank commander may be justified in making a range change.)

93. Subsequent Fire Commands

Subsequent fire commands are given by the tank commander only when the gunner fails to observe a round through his direct-fire sight or when the tank commander desires to take over the adjustment of fire. When subsequent fire commands are necessary, they are issued in the following sequence:

a. Stationary Targets.

(1) Deflection change (in mils).

(2) Range change (in yards).

(3) Command to fire.

b. *Moving Targets.*
   (1) Range change (in yards).
   (2) Lead change (in leads).
   (3) Command to fire.

c. *To Change Ammunition or Fuze.* When firing, it may be necessary to designate a different type of ammunition. For example, if a round of SHOT has penetrated a pillbox or heavy masonry building and the tank commander decides to fire HE through the opening, he commands FIRE HE. This alerts all crewmen to a change in the ammunition. The loader at once loads the HE round and continues to load HE until he hears CEASE FIRE or another change. The gunner, hearing the change in ammunition, indexes the correct (new) ammunition code number. The commander uses the same procedure to change the fuze setting; for example, to change from fuze superquick to fuze delay, he commands FIRE FUZE DELAY. Normally, a chambered round will be fired even though a new type ammunition has been designated.

Note. Elements may be omitted from the subsequent fire command if not applicable to an adjustment; therefore the subsequent fire command may contain one, two, or three elements.

94. Moving Targets

a. *General.*
   (1) A moving target is one which has apparent speed (app. III, FM 17–12). Targets moving across the line of sight either horizontally or diagonally have apparent speed. Targets moving directly toward or directly away from the tank have no apparent speed.
   (2) The gunner will range on the target in all cases except when engaging dangerous surprise targets or rapidly moving targets, at which time the unit battle sight will be used.

b. *Leading.* When taking moving targets under fire, the gun must be aimed in front of the target to cause the projectile and target to meet. This technique is called "leading." A lead is 5 mils and is taken from the center of the target. The proper lead is maintained by tracking.

c. *Tracking.* In order to maintain the proper lead, the gunner must cause the movement of the gun to keep pace with the movement of the target. This technique is called "tracking." While the tank commander announces his initial fire command, he lays the gun for direction with his power control handle and M20 periscope. The gunner indexes the code number for the type ammunition announced. The tank commander continues to track the target with no lead until the gunner announces IDENTIFIED. The gunner then takes over control of the turret and tracks with no lead while ranging. Upon completion of ranging, the gunner swings through the target and tracks with the prescribed lead. In tracking, the gun should be traversed "through the target" from behind to ahead of the center of target until the proper lead is applied. The gunner will not stop the movement of the gun while he fires; nor will he attempt to ambush the target by moving ahead of it, then stopping, and firing when the target reaches the proper lead on the sight reticle. He will track with a smooth, continuous motion, and will maintain a constant sight picture. He will track continuously before, during, and after firing. If he misses, he will then make necessary changes in elevation and lead to obtain a target hit. Tracking is a combination of traversing and changing elevation in order to maintain a proper sight alinement.

d. Initial Fire Commands. Initial fire commands are the same as those used for engaging stationary targets, except that the lead element will be announced just before the command to fire (par. 91). Normally, one lead will be used initially regardless of target speed or range. For example, the tank commander may give the following initial fire command: GUNNER, SHOT, TANK, ONE LEAD, FIRE.

e. Sensings. Rounds fired at moving targets are sensed in relation to the target as when firing at stationary targets. Any deflection errors are lead errors, and the actual mil error must be converted to leads and fractions of leads.

f. Adjustment. The burst-on-target method will be used unless the gunner cannot apply this method and announces LOST, at which time the alternate method will be used. The tank commander will mentally sense each round fired and be prepared to announce his sensing and issue a subsequent fire command.

g. Subsequent Fire Commands. When necessary, the tank commander will issue subsequent fire commands. The lead correction will be announced as a change in leads rather than in mils. For example, if a round passes behind the aiming point, the tank commander will announce ONE MORE or ONE-HALF MORE. The gunner will then increase his lead accordingly. Conversely, if the round passes in front of the aiming point, the tank commander will announce ONE LESS or ONE-HALF LESS. Range corrections will be announced when necessary, in the same manner as when adjusting on stationary targets.

Section II. ADJUSTMENT OF FIRE

95. Adjustment of Fire, Primary Method

The primary method of adjustment is burst-on-target, in which the gunner, observing through his direct-fire sight, notes the point on the sight reticle where the burst or tracer appears in relation to the target and, without command from the tank commander, moves that point of the gun-laying reticle onto the center of the target for subsequent rounds. This method of adjustment provides a quick, accurate means of obtaining second-round target hits. The gunner uses this method whenever possible. Typical example of burst-on-target adjustment using the M12 range finder reticle as well as the M20 periscope reticle appear below:

a. *Situation 1.* M47 tank mounting a 90-mm gun. Target: an antitank gun.

   (1) The tank commander gives the following fire command: GUNNER, HE, ANTITANK, FIRE. He lays the gun for direction with his power control handle and M20 periscope.

   (2) The gunner, upon viewing the target, announces IDENTIFIED and ranges on the target (fig. 40).

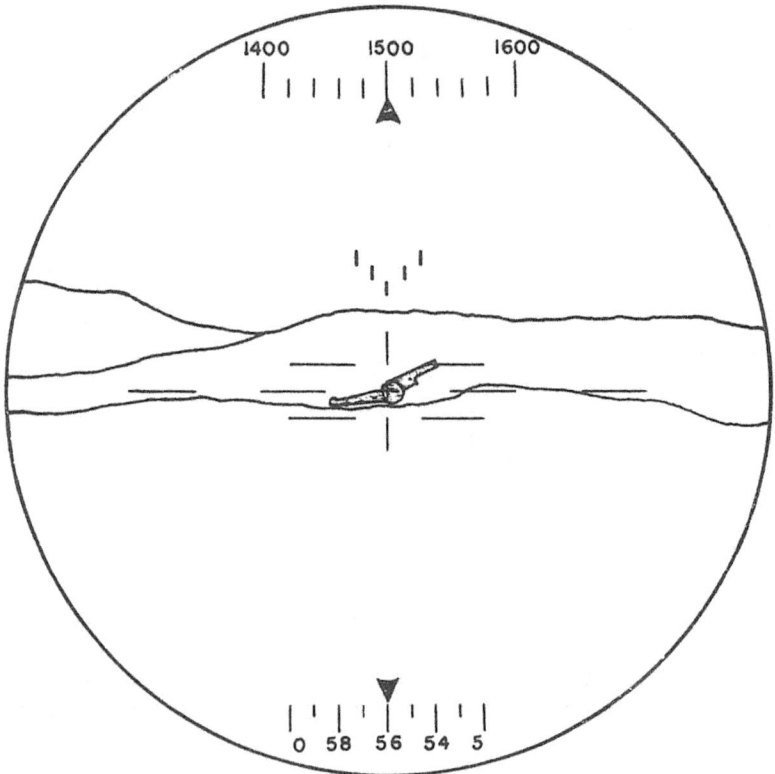

Figure 40.

   (3) As soon as the gunner determines the range, he places the aiming cross on the center of the target, announces ON THE WAY, pauses one second, and fires (fig. 41).

   (4) The round fired in figure 41 was off in deflection to the right. The gunner notes that point on the sight reticle and, with the turret controls, moves that point to the center of the target (fig. 42) without command, fires, and obtains a target hit.

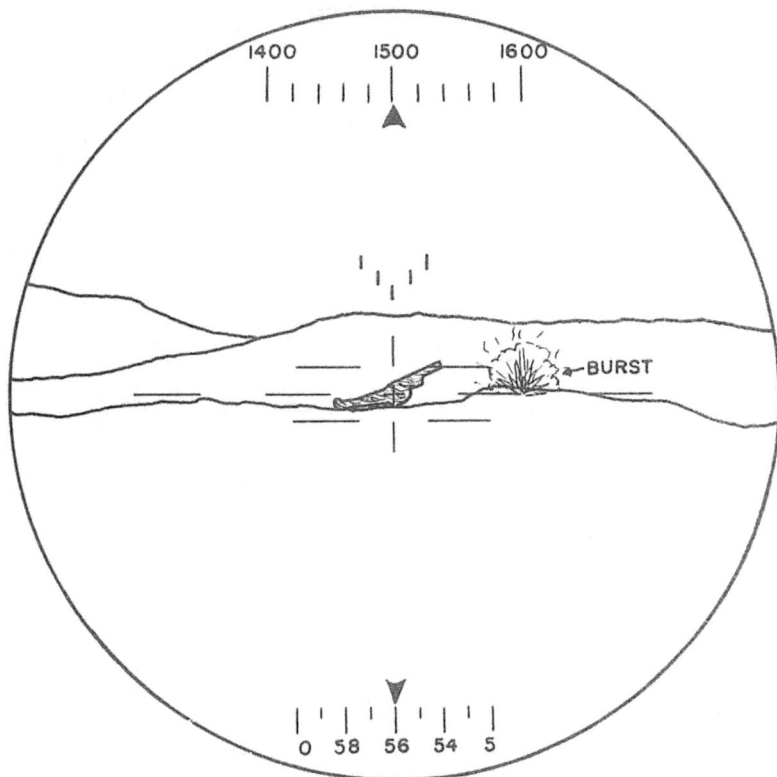

Figure 41.

b. *Situation 2.* M47 tank mounting 90-mm gun. Target: a stationary tank. The range finder has become inoperative and the M20 periscope is being used.

    (1) The tank commander gives the following fire command: GUNNER, SHOT, 800, TANK, FIRE. He lays the gun for direction with his power control handle and M20 periscope.

    (2) The gunner indexes the announced range of 800 yards for AP ammunition on the range drum. He lays the aiming cross of the periscope on the center of the target, announces ON THE WAY, pauses one second, and fires (fig. 43).

    (3) The gunner notes that the tracer of the round fired in figure 43 passed over the target to the left. The gunner notes the point on the sight reticle where the tracer appeared as it passed the target, moves that point onto the center of the target (fig. 44) without command, fires the next round, and obtains a target hit.

c. *Situation 3.* M47 tank mounting 90-mm gun. Target: a moving tank. One of the range finder end boxes has been damaged and the gunner is using the range finder as an offset telescope.

    (1) The tank commander gives the following fire command: GUNNER, SHOT, 900, TANK, ONE LEAD, FIRE. He lays the gun for direction with his power control handle and M20 periscope.

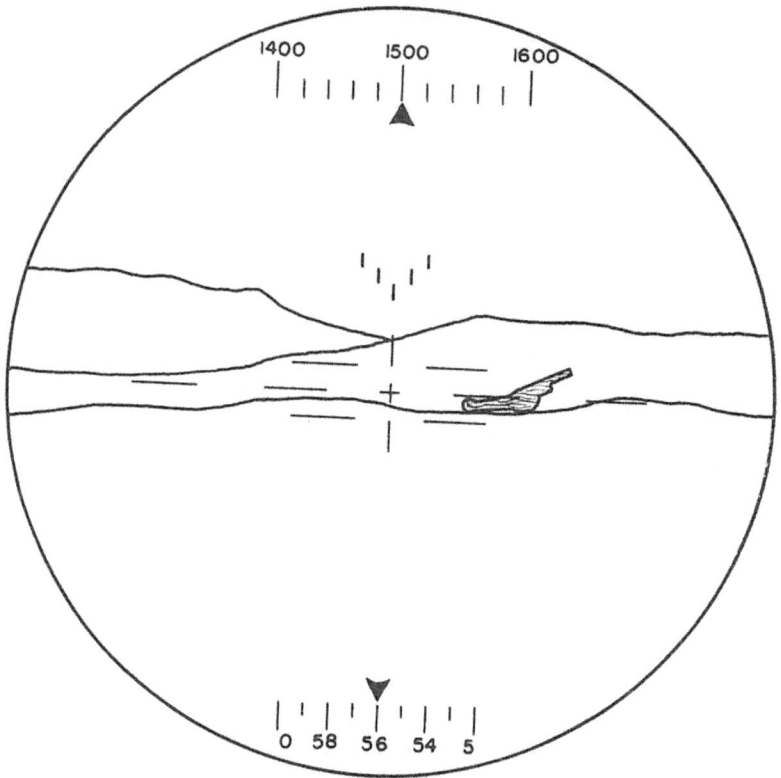

Figure 42.

(2) The gunner indexes the proper ammunition code number on the ammunition scale and the announced range on the range scale. After viewing the target and announcing IDENTIFIED, the gunner takes one lead from the center of the target, announces ON THE WAY, pauses one second, and fires (fig. 45).

(3) The gunner notes that the tracer (fig. 45) passed above and behind the center of the tank. He notes the point on the sight reticle at which the tracer appeared as it passed the tank, moves that point onto the center of the tank (fig. 46), fires the next round, and obtains a target hit.

96. Adjustment of Fire, Alternate Method

The alternate method of adjustment is utilized by the tank commander when the primary method cannot be used effectively due to obscuration, terrain, or extreme range. The alternate method of adjustment involves the use of standard range changes to be announced by the tank commander under certain conditions. The conditions and standard range changes are as follows:

a. When the range finder has been used to determine the initial range to the target and the gunner, upon firing the initial round, fails to ob-

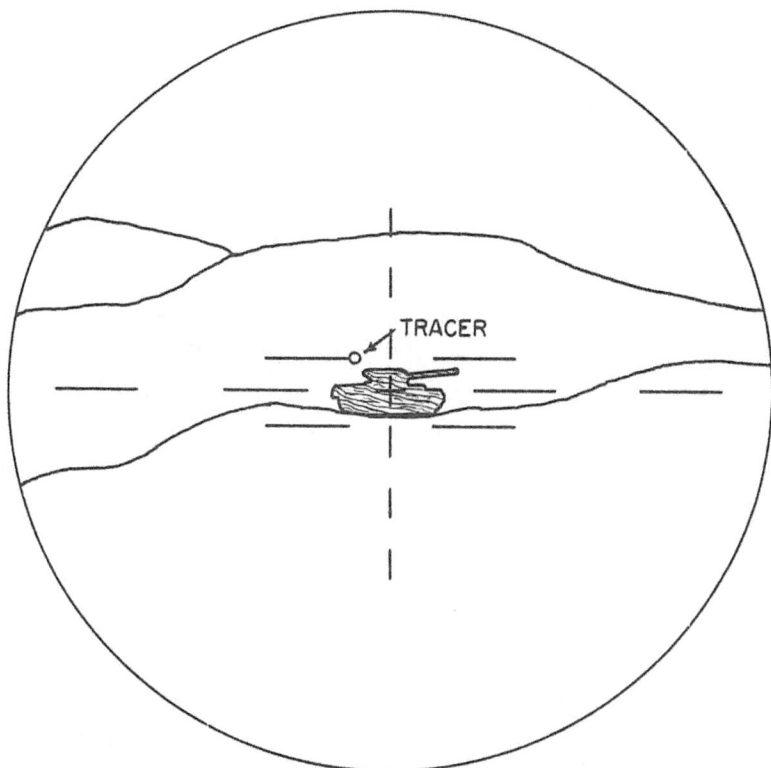

Figure 43.

serve the tracer or burst in his sight, he will announce LOST. The tank commander will then announce a sensing and subsequent fire command, adding or dropping 200 yards if there was a range error, regardless of range to the target. The gunner, upon seeing this round in his direct-fire sight, will apply burst-on-target to obtain a target hit; however, if this second round is also LOST to the gunner, the tank commander will continue with the adjustment, making whatever deflection and range changes he feels are necessary to obtain a target hit. (Deflection errors are measured with the binocular, and range changes are made in multiples of 50 yards.) If the necessary range change is less than 50 yards, the command may be ADD (DROP) A HAIR.

Note. For practical purposes a 1-mil elevation change will change the range 100 yards. During the alternate method of adjustment, the gunner will use the range lines of the gun laying reticle to make the necessary range change.

   b. When the initial range to the target is estimated by the tank commander and the gunner fails to see the tracer or burst, he will announce

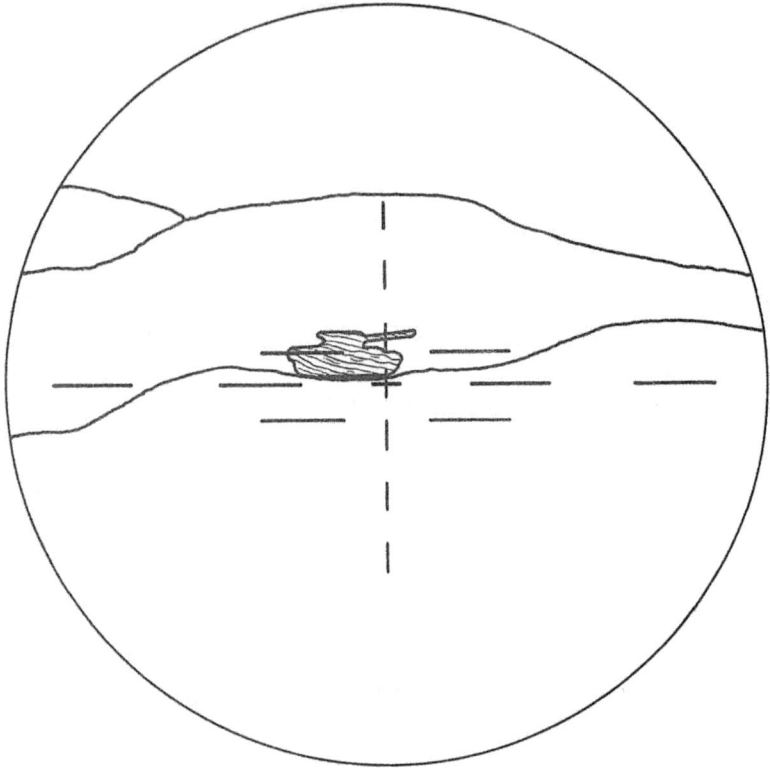

Figure 44.

LOST. The tank commander will then announce a sensing and subsequent fire command, making a range change as follows:

    (1) If the estimated range to the target is *1500 yards or less*, he will add or drop 200 yards of range for the first subsequent round. The subsequent adjustments are the same as listed in *a* above.

    (2) If the estimated range to the target is *over 1500 yards*, he will add or drop 400 yards for the first subsequent round. Again, if the gunner observes this round, he will apply burst-on-target; if the round is LOST to the gunner, the tank commander will continue with the adjustment as described in *a* above.

    Note. If an extremely large error is made in the initial *estimated* range, the tank commander may announce a new range.

    c. When the gunner, during an adjustment, fails to observe a round *after* applying the burst-on-target method, he will announce LOST.

144

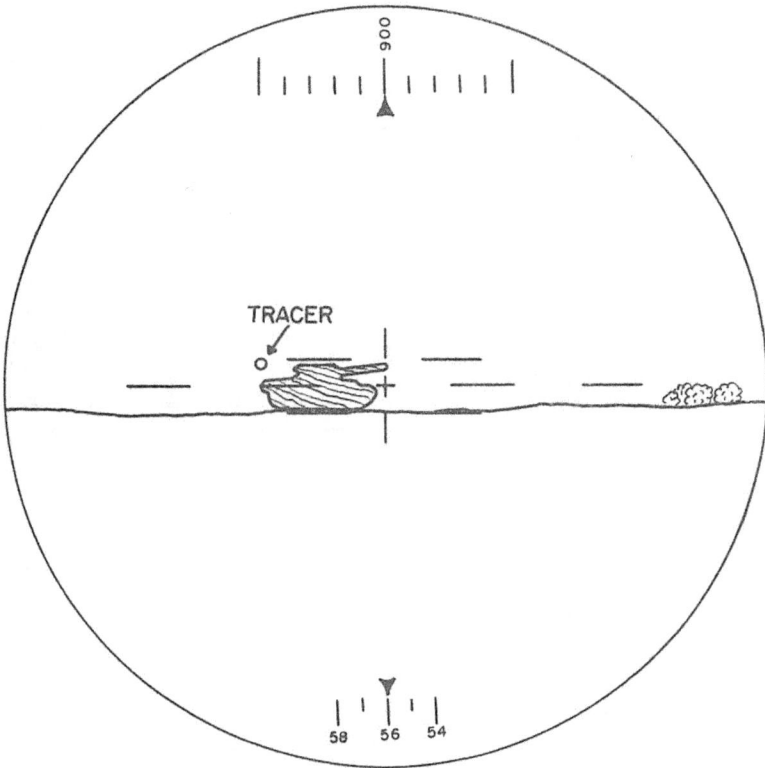

Figure 45.

The tank commander will take over the adjustment and use the alternate method of adjustment, making the deflection and range changes he feels are necessary to obtain a target hit.

d. When the gunner and tank commander both fail to observe the round, the gunner will announce LOST. The tank commander will take over the adjustment, announcing LOST and giving a subsequent fire command. He may fire another round without changing the range, or he may give a range change to bring the next round to where it can be observed.

e. In all of the above cases, the gunner will apply the announced range change by use of his direct-fire sight and will use the primary method of adjustment whenever possible.

Note. The tank commander maintains control of his tank at all times and may take over adjustment of fire at any time. Once the initial range has been changed or a target hit has been obtained, the standard range change is not necessary.

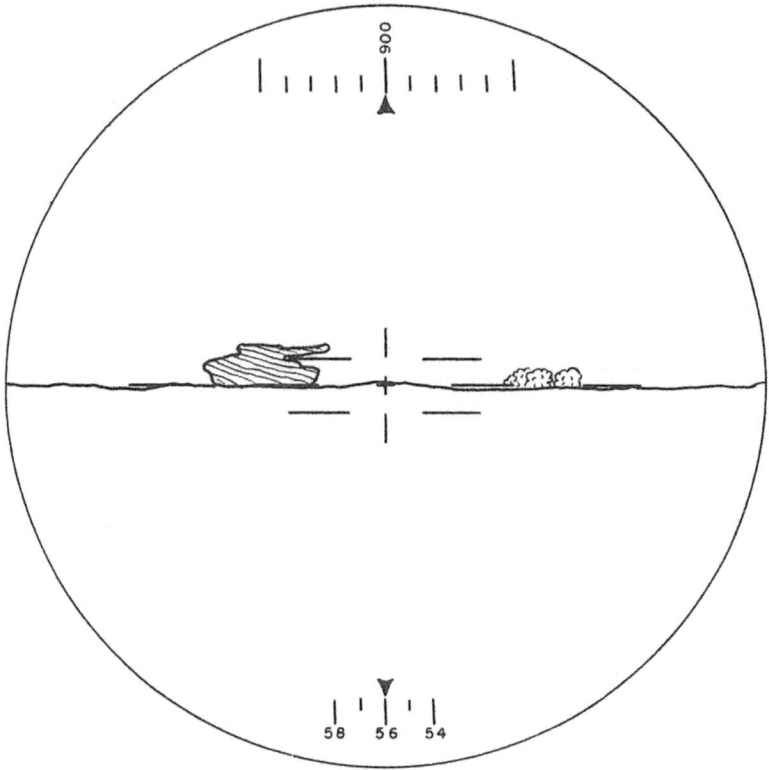
Figure 46.

97. Firing Tank Machineguns

Proper utilization of tank-mounted machineguns adds to the shock action of the tank and provides an effective means of destroying hostile personnel and unarmored vehicles. The M47 tank has three machineguns: the coaxial machinegun, the turret-mounted machinegun, and the bow-mounted machinegun. The technique involved in firing these weapons is as follows:

  a. *Coaxial Machinegun.*

    (1) The tank commander issues an initial fire command, following the sequence listed in paragraph 91, and lays the gun for direction using the power control and M20 periscope. He announces a range element. The gunner will not range on targets to be engaged with the coaxial machinegun.

146

(2) The gunner fires the coaxial machinegun, utilizing the range finder or M20 periscope sight, the elevating and traversing controls, the gun selector switch, and the firing triggers. The gunner, upon command, fires the coaxial machinegun in bursts of 20 to 25 rounds, observing the tracer stream through his direct-fire sight. By manipulating the gun, he moves the tracer stream onto the target and continues to fire (traverses and searches the target area if necessary) until the tank commander commands CEASE FIRE.

b. Turret-Mounted Machinegun.

(1) This machinegun is mounted on a pedestal mount or in the cupola on the top of the turret and is fired by either the tank commander or loader (depending on type of mount) at ground and aerial targets.

(2) No fire commands are involved in firing this machinegun.

(3) Against ground targets, the turret-mounted machinegun is fired in bursts of 10 to 20 rounds. Tracer observation is the method used. The tank commander or loader will engage targets of opportunity or fire as the tank commander directs.

(4) When engaging aerial targets, the turret-mounted machinegun is fired in one continuous burst. By observing the tracers, the tank commander or loader manipulates the tracer stream onto the target by tracking and leading.

c. Bow-Mounted Machinegun.

(1) The bow gunner fires the bow-mounted machinegun on targets of opportunity and as the tank commander directs.

(2) The tank commander issues fire commands to the bow gunner when he desires to engage targets. These fire commands are issued in the following sequence:

(*a*) Alert_____ BOG
(*b*) Direction_____ FRONT
(*c*) Description_____ TROOPS
(*d*) Command to fire_____ FIRE

Note. The range may be added, if needed, to clarify the target location.

(3) The bow gunner fires the bow-mounted machinegun in bursts of 20 to 25 rounds, observing the tracer stream through his vision device. By manipulating the bow machinegun in the flexible ball mount, the bow gunner moves the tracer stream onto the target, and then traverses and searches the target area, continuing to fire in bursts of 20 to 25 rounds until the tank commander commands CEASE FIRE.

98. Special Conditions

Conditions arise that prevent the use of normal techniques of fire, requiring that substitute means be utilized to insure destruction of targets. A condition that commonly arises is one in which the gunner cannot see the target through his direct-fire sight—for example, when the tank is in a defiladed position, during night firing, or when obscuration exists. In this event, the tank commander issues a six-element initial fire command in the sequence shown in paragraph 91. The range element is announced in hundreds of yards; and the gunner, referring to an aiming data chart, ballistic unit, or firing table, determines the elevation for the announced range. He applies this to the M13 elevation quadrant, centering the bubble by elevating or depressing the gun. The directional element of the fire command may be given by the reference point method, in which the tank commander lays the gun on a reference point and has the gunner traverse right or left, a given number of mils, using the azimuth indicator. The tank commander observes the effect of the fire and uses the alternate method of adjustment to obtain target hits. Another special condition that requires consideration is the adjustment of ricochet fire. The gunner cannot apply the primary method of adjustment when firing ricochet fire, therefore, the tank commander senses the effects of the fragments upon the ground and utilizes the alternate method of adjustment to bring that effect upon the target. For a more detailed discussion of special conditions, see FM 17–12.

Section III. TANK GUNNERY QUALIFICATION COURSE

99. Purpose and Scope of Course

a. The purpose of the tank gunnery qualification courses is to provide a means of determining the proficiency of the tank crewman in gunnery. The courses are designed both to test the gunner and to serve as an adjunct to training in the proper care and use of the weapons and their accessories. The tables of the courses will be fired for training purposes before they are fired for record. The tank gunnery qualification courses cover the gunner's preliminary examination, subcaliber firing exercises and service firing exercises. The courses are—

    (1) Standard—tables I through VIII.

    (2) Limited—tables I through IV.

b. Each tank crewman must pass the Gunner's Preliminary Examination with a score of 80 percent or higher before firing either the Standard or Limited Course. Gunners firing the Standard Course must attain a score of 280 or higher over tables I through IV before firing tables V through VIII. Since tank crewmen of the active Army must be prepared for combat without further training, a limited qualification and classification for such personnel is not acceptable. Therefore, use of the Limited Course for classification purposes is restricted to tank crewmen of the Reserve components. When ranges are not available to active Army personnel for service firing (tables V, VI, VII and VIII), the full Limited Course, practice and record firing, may be fired twice annually for practice; however, classification of gunners will not be made.

c. In order to obtain maximum use of facilities, tables may be fired in any order within each group except that table I will be fired first in the subcaliber exercises, and table V will be fired first in the service firing exercises.

100. Classification of Gunners

a. The total possible points that can be scored in each phase of qualification are as follows:

	Possible points
Gunner's Preliminary Examination	320
Subcaliber firing exercises (tables I–IV)	400
Service firing exercises (tables V–VIII)	400

In order to obtain maximum use of facilities, tables may be fired in any order within each group except that table I will be fired first in the subcaliber exercises, and table V will be fired first in the service firing exercises.

b. The required score for each gunner classification is as follows:

Classification	Score
STANDARD COURSE	
Expert Gunner	720–800
First-Class Gunner	640–719
Second-Class Gunner	560–639
Unqualified	559 or less
LIMITED COURSE	
Limited First-Class Gunner	360–400
Limited Second-Class Gunner	320–359
Unqualified	319 or less

101. Ammunition Required

The following table lists the ammunition required to fire a Tank Gunnery Qualification Course.

Course (Fired once for practice and once for record)	Ammunition				
	Coaxial Machinegun			Service HE	Service SHOT
	Tracer	Ball	Frangible		
Table I [1]	30				
II [2]	15				
III	30	120			
IV [2][3]			15		
V					4
VI				4	4
VII					10
VIII				1	
Total for practice	75	120	15	5	18
Total for record	75	120	15	5	18

[1] Includes 10 rounds for zeroing and failure to fire single shot.
[2] Includes 3 rounds for zeroing.
[3] Tracer ammunition may be used if frangible ammunition is not available.

102. Rules for Record Firing

a. Before record firing, the gunner is required to check the condition of the weapon, sights, and ammunition. He is permitted a reasonable length of time to do this.

b. Only the examining personnel, and the assigned crew of which the gunner being tested is a member, will be in the tank during record firing. The examining officer will take his position on the inside of the tank whenever possible.

c. During firing, the gunner performs all the operations required by the test, without the benefit of coaching or assistance. Other members of the crew perform their normal duties. The assigned crewmen will rotate positions within their tank when firing the Tank Gunnery Qualification Course.

d. All exercises will be fired with the sights adjusted by boresighting and zeroing. (The established zero will be used for tables VI, VII, and VIII.)

e. If a misfire or other malfunction of the main gun occurs, the gunner will announce MISFIRE; if the machinegun fails to fire, the gunner will announce STOPPAGE. Thereafter, no one is allowed to touch the gun without authorization of the examining officer. The examining officer notes the time and examines the gun. After the malfunction has been corrected, the crewmen will be permitted to complete the exercise. Unless the malfunction was due to the negligence of the gunner, the time it takes to correct the malfunction will not be counted against the time allowed the gunner for that particular phase of the test.

f. Prior to record firing, the examining officer must thoroughly familiarize himself with his duties and the correct firing procedure.

103. Gunner's Preliminary Examination, General

The Gunner's Preliminary Examination will be conducted by the company commander and such officer and enlisted assistants as may be necessary. The examination will be given to each member of the tank crew, and a score of 80 percent or more will be required before the crewman is permitted to fire the tables of the Tank Gunnery Qualification Course.

104. Materiel Examination

a. Disassembly of Breechblock. The breech cover is removed and the gun traveling lock disengaged. The examining officer commands, DISASSEMBLE BREECHBLOCK, and starts his timing with a stop watch. The gunner is required to remove and disassemble the breechblock, using the prescribed method. If the disassembly is completed in 3 minutes or less, a credit of 20 points is given. A cut of 5 points is made for every additional 30 seconds or part thereof when more than 3 minutes is taken for this test. To facilitate the handling of parts in disassembly and assembly of the breechblock, all parts should be free from oil and grease. A tarpaulin or shelter half should be placed on the floor of the turret to avoid losing or damaging parts, and the turret lights should be turned on to improve visibility. The examining officer will note any particular difficulty encountered by the gunner in removing or replacing any part. If the difficulty is due to the part being bent, burred, or otherwise damaged, the examining officer will note the time required to remove or replace that part and deduct it from the total time required. The closing spring will *not* be removed in this test.

b. Assembly of Breechblock. The examining officer commands, ASSEMBLE BREECHBLOCK, and starts the timing with the stop watch. The gunner is required to assemble and replace the breechblock, using the prescribed procedure. If the assembly and replacement is correctly performed in 4 minutes or less, a credit of 20 points is given. A cut of 5 points is made for every additional 30 seconds or part thereof when more than 4 minutes is taken for this test. One other crewman (preferably an instructor) may assist the gunner when appropriate.

c. Care and Maintenance. The examining officer commands, PERFORM DAILY LUBRICATION CHECK, INCLUDING CHECKING AND FILLING OF RECOIL SYSTEM. The gunner points out the lubricating points for which he is responsible and, from a display of lubricants and lubricating devices, selects the proper ones to be used for each point. All lubricants will be in the containers in which they are normally issued. The gunner also explains the procedure for checking, filling, and bleeding the recoil system. The total possible credit is 20 points. A penalty of 4 points is assessed for each of the following errors:

    (1) Each point of lubrication missed.

    (2) Each error in selection of lubricating device or lubricant.

    (3) Each error in procedure for checking, filling, and bleeding the recoil system.

d. Sight Adjustment. The vehicle is placed in a position where several features or objects, suitable for sight adjustment, are in view. These objects should be at varied known ranges of from 500 to 3000 yards. The gunner must select a target as near 1500 yards as possible for boresighting. The examining officer places the Range Finder, M12, out of adjustment in any or all of the following: ammunition setting, halving, rheostat setting, range scale, interpupillary or diopter setting, or ICS setting. The secondary equipment items, consisting of the ballistic unit and the gunner's Periscope, M20, are also placed out of adjustment. The examining officer commands, MAKE BORESIGHT ADJUSTMENT. The gunner is required to adjust the range finder and periscope, using the prescribed methods. The total possible credit is 50 points. A score of 30 points is given for placing the range finder in proper adjustment, and 20 points for placing the periscope in proper adjustment. No partial credit will be given for sights which are not in proper adjustment.

e. Putting Turret Into Power Operation. The examining officer commands, PUT TURRET INTO POWER. The gunner is required to put the power traverse mechanism into operation, performing all steps in the prescribed sequence. The total possible credit is 10 points. A penalty of 5 points is assessed for each error.

Note. The following steps are performed in putting the turret into power operation. Remember the word "ACUTE" for a key to the steps.

A—Alert crew_____ Insure that crew is in safe position; check immediate area for obstructions.

C—Check oil_____ Oil should be at "FULL" mark on bayonet gage.

U—Unlock turret_____ Traverse manually to make sure the turret is unlocked. Return manual traverse control handle to latched position over the dump valve micro switch.

T—Turn on power_____ Power control handles should be in the neutral position and the loader's traverse safety on.

E—Elevate and traverse_____ Operate in power to make sure controls are functioning properly.

f. Testing Gunner's Quadrant. The examining officer commands, TEST GUNNER'S QUADRANT. The gunner is required to make the end-for-end test on a quadrant which is out of adjustment. The gunner is also required to explain the method of determining the correction. The total possible credit is 10 points. A penalty of 5 points is assessed for each error.

g. Adjusting Elevation Quadrant. The elevation scale and the micrometer scale of the elevation quadrant are placed out of adjustment. The examining officer commands, ADJUST ELEVATION QUADRANT. The gunner is required to adjust both the elevation and the micrometer scales, using the prescribed procedure. No credit is given if the adjustments are not precise. The total possible credit is 10 points.

h. Identification and Inspection of Ammunition. All standard types of ammunition, for both the tank gun and the machineguns, should be displayed. Some rounds should have apparent faults, such as a dented cartridge case, badly burred rotating band, or adhering dirt. Lettering and markings on service ammunition are covered. The gunner is required to identify five rounds of ammunition as they are pointed out to him. Color, shape, and size are used as means of identification. The gunner is also required to locate the defects in three rounds that are pointed out to him, and to describe the results of using these rounds without correcting the defect. The total possible credit is 10 points. A penalty of 2 points is assessed for each error.

105. Simulated Firing Examination

a. Direct Laying, Range Finder, M12. In this test the gunner, using the range finder, is required to range and lay the gun on four targets.

(1) The tank is placed in a position from which five or more targets can be seen. These targets must be at ranges known only to the examining officer. The gunner places the range finder in operation, sets the unit battle sight, and announces READY.

(2) The examiner gives a four-element (ammunition element of which is preset battle sight) fire command; he will also lay the gun for direction using the tank commander's power control handle and M20 periscope. (The aiming point on the target should be well defined to eliminate confusion as to correct laying of the gun.) The examiner starts timing and the gunner starts the test as soon as the examiner lays the gun for direction. The gunner ranges, makes final lay of the gun on the target, and announces ON THE WAY. The examiner will check the sight picture and range scale setting. The gunner will be allowed four UOE (Units of Error) in his rangings. If he ranges within the allowable

four UOE and has the proper sight picture within 10 seconds, he is given 10 points credit (5 points for correct range and 5 points for correct sight picture). No credit is given if his time for completing the trial exceeds 10 seconds.

(3) The test is then repeated three times, using the procedure in (2) above, for a total of four trials.

(4) The total possible score for this test is 40 points.

b. *Direct Laying, Periscope, M20, and Ballistic Unit.* In this test the gunner sets off the correct range on the ballistic unit for the ammunition announced by the examining officer, and lays the gun for the correct sight picture.

(1) The tank is placed in a position from which several targets can be seen. The gunner checks the diopter setting on the M20 periscope, sets the unit battle sight on the ballistic unit, and announces READY.

(2) The examiner gives a five-element initial fire command while laying the gun for direction with the commander's power control handle and M20 periscope. (The aiming point on the target should be well defined to eliminate confusion as to correct laying of the gun.) The examiner starts timing and the gunner starts the test (sets range on ballistic unit) when the range is announced. The gunner makes final lay of the gun on the target and announces ON THE WAY. The examining officer checks the initial sight picture, using the commander's M20 periscope, and checks the ballistic unit to determine whether the correct range has been set opposite the announced ammunition. If the gunner performs this exercise correctly within 10 seconds, he receives 10 points. He will receive no credit if he does not perform the exercise correctly within 10 seconds.

(3) The examining officer then gives a subsequent fire command, using a deflection change of not more than 10 mils and a range change of not more than 400 yards, as prescribed in the alternate method of adjustment. The gunner lays the gun with the corrected sight picture and announces ON THE WAY. The examining officer checks the sight picture, using the commander's M20 periscope. If the gunner performs this exercise correctly within 5 seconds, he receives 10 points. He will receive no credit unless he has the correct sight picture within 5 seconds.

(4) The examining officer then gives a second subsequent command, using the procedure in (3) above.

(5) The total possible score for this test is 30 points.

c. *Use of Elevation Quadrant.* In this test the gunner lays the gun for range, using the elevation quadrant.

(1) The examining officer announces a range in yards. Using the ballistic unit as an aiming data chart, the gunner determines the elevation corresponding to the announced range, insures that the ballistic unit is returned to zero, applies this elevation to the elevation quadrant, and centers the bubble, using the turret controls. He then announces ON THE WAY. The examining officer checks the elevation setting and the correct lay of the gun.

(2) The examining officer then announces two subsequent ranges, and the same procedure is followed.

(3) No credit will be given if an improper quadrant setting is used or if the bubble is not accurately centered. For each trial, if the trial is correctly performed in exactly 9 seconds or less, the gunner receives 10 points. He will receive no credit if the time exceeds 9 seconds. The total possible score for this test is 30 points.

d. Use of Gunner's Quadrant, M1. In this test the gunner sets an elevation on the M1 quadrant and places the quadrant properly on the quadrant seats.

(1) The gunner sets the M1 quadrant at zero. The examining officer announces an elevation. The gunner applies this elevation, properly seats the quadrant on the breech, centers the bubble, and announces SET. The examining officer checks the quadrant elevation and the position of the bubble.

(2) The examining officer then announces two subequent elevations, one of which will be to an even tenth of a mil, and the same procedure is followed.

(3) No credit will be given if the elevation is incorrect or if the quadrant is seated improperly. For each trial, if the trial is correctly performed in exactly 7 seconds or less, the gunner receives 10 points. He will receive no credit if the time exceeds 7 seconds. The total possible score for this test is 30 points.

e. Adjustment Using Azimuth Indicator and Elevation Quadrant.

(1) The examining officer sets the azimuth indicator at zero and the elevation quadrant at an arbitrary elevation, centering the bubble with the turret controls. This takes the place of laying the gun from an initial fire command.

(2) The examining officer then gives four subsequent commands. The deflection shifts should be in multiples of 5 mils and should not exceed 100 mils. The gunner follows the commands, using the azimuth indicator for deflection changes and the elevation quadrant for range changes. After each azimuth indicator and quadrant setting has been applied and the bubble centered, the gunner announces ON THE WAY.

The examining officer checks the azimuth indicator and quadrant setting and the bubble of the elevation quadrant.

(3) No credit will be given if any setting is incorrect or if the bubble of the quadrant is not centered. For each trial, if the trial is correctly performed in exactly 20 seconds or less, the gunner receives 10 points. He will receive no credit if the time exceeds 20 seconds. The total possible score for this test is 40 points.

106. Table of Possible Scores

Gunner's Preliminary Examination	Possible Points
Materiel Examination	
Disassembly of breech mechanism	20
Assembly of breech mechanism	20
Care and maintenance	20
Sight adjustment	50
Putting turret into power operation	10
Testing gunner's quadrant	10
Adjusting elevation quadrant	10
Identification and inspection of ammunition	10
Total possible score	150
Simulated Firing Examination	
Direct laying, primary sighting devices	40
Direct laying, secondary sighting devices	30
Use of elevation quadrant	30
Use of gunner's quadrant	30
Adjustment using azimuth indicator and elevation quadrant	40
Total possible score	170
TOTAL POSSIBLE SCORE, GUNNER'S PRELIMINARY EXAMINATION	320

Note. A gunner must score 256 or more points to be eligible to fire subcaliber or service exercises.

107. Zeroing for Subcaliber Firing

a. Cover the right end box of the range finder.

b. Loosen the two cap screws which secure the front mounting bracket of the coaxial machinegun.

c. Install the coaxial machinegun in its mount.

d. Loosen the two socket head screws that lock the elevation adjustment and the socket head screw that locks the traverse adjustment.

e. Center the front barrel bearing of the machinegun in the aperture of the gun mantlet by use of the elevation adjustment.

f. Traverse the receiver of the machinegun as far to the right (toward the main armament) as it will go.

g. Grasp the barrel of the machinegun and pull it to the left until the front barrel bearing barely touches the left side of the aperture in the gun mantlet. Tighten the cap screws which secure the front mounting bracket of the machinegun. Tighten the socket head screws which secure the elevating and traversing mechanism.

h. Boresight in the following manner:
    (1) Remove back plate, bolt handle, and bolt from the machinegun.
    (2) One person sights through the bore while another manipulates the turret and gun controls to lay on a definite point at 1000 inches.
    (3) Index 90 on the ammunition scale of the M12 range finder.
    (4) Unlock the elevation and azimuth boresight knobs and lay the aiming cross on the aiming point referred to in (2) above. Relock the knobs.

> *Note.* If the elevation boresight knob will not move the reticle a sufficient amount, the ballistic correction knob may be used to finish the vertical adjustment.

i. Zeroing is accomplished in the following manner:
    (1) Lay on a definite aiming point at 1000 inches.
    (2) Fire one or more rounds single shot. Re-lay after each round.
    (3) Without disturbing the lay of the gun, refer the sights to the center of the shot group by use of the elevation and azimuth boresight knobs.
    (4) Fire one or more rounds to confirm the zero.

108. Subcaliber Firing Exercises

All subcaliber firing will be conducted with the coaxial machinegun. When single shots must be fired, ammunition should be loaded with alternate dummy rounds, or a single-shot device may be used. CONARC-Approved Training Aid No. 1, Single-Shot Device, may be used; plans for this aid are available at Training Aids Subcenters. Targets will be *physically* scored during all record firing.

a. Table I: Subcaliber Manipulation Exercise, 1000-Inch.
    (1) The purpose of this exercise is to test the gunner's ability to manipulate the turret controls and fire at stationary targets prior to firing service ammunition.
    (2) This subcaliber gunner's exercise in manipulation requires the gunner to fire rapidly on a series of stationary targets; figure 47 illustrates the target layout. The 4 x 4-inch targets shown in the illustration may be stapled or tacked onto staves.

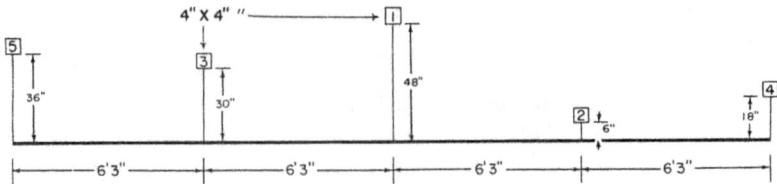

Figure 47. Manipulation target layout.

Note. The right end box of the range finder is covered, "B" is indexed on the range scale, and the coaxial machinegun is zeroed for 1000-inch firing. This also applies to the firing of the exercise in (3) and (4) below.

(3) The gunner lays on the left (No. 5) target. At the command, COMMENCE FIRING, he lays the aiming cross of the range finder gun laying reticle on the center (No. 1) target and fires one round. He then fires one round at each of the remaining four targets in the order in which they are numbered. Time is recorded from the command, COMMENCE FIRING. At the end of the allowed time, the examining officer will command, CEASE FIRING. Rounds fired after this command will be scored as misses.

(4) The exercise consists of four trials in (3) above, two in manual traverse and two in power traverse. Credit will be given in accordance with table I, below.

(5) The exercise will be scored as follows:

Table I

(Possible score: 100 points)

Trials	Number rounds	Time (Seconds)	Points each hit
1st—Manual traverse	5	30	5
2d—Manual traverse	5	30	5
3d—Power traverse	5	40	5
4th—Power traverse	5	40	5

Note. Targets will be scored and marked, and replaced when required.

(6) See figure 48 for illustration of an appropriate score card.

b. Table II: Subcaliber Shot Adjustment, Moving Target Exercise, 200 feet.

(1) To purpose of this exercise is to test the gunner's ability to track, fire on, and adjust fire on moving targets prior to firing service ammunition.

(2) In this exercise, the gunner is required to fire tracer (single shot) at moving targets on a 200-foot range, using the coaxial machinegun.

158

Co ___A___

Bn ___1ST___

NAME *SMITH JOHN J.*

RANK *PFC* SN *38132977*

DATE ___/ FEB 54___

70

Total Score

TANK GUNNERY QUALIFICATION COURSE SCORE CARD

100 points
Possible

TABLE I (SUBCALIBER MANIPULATION EXERCISE—1000-INCH)

TRIALS	NUMBER OF ROUNDS	POSSIBLE	MAX TIME (SEC)	TARGET HITS					SCORE
				1st RD	2d RD	3d RD	4th RD	5th RD	
1st—Manual	5	25	30	5	5	5	0	0	15
2d—Manual	5	25	30	5	5	0	5	5	20
3d—Power	5	25	40	5	5	5	5	0	20
4th—Power	5	25	40	5	5	0	5	5	20
							TOTAL SCORE		75

A. Five points for each target hit. No credit for hits obtained after time limit.

B. Add score column to obtain total score for this exercise.

Satisfactory Score _____ 70 points.

Lt Leonidas R. James

Examining Officer's Signature

Figure 48. Score card for table I.

(3) Targets are mounted on a 6 x 6-foot panel as shown in figure 49. The target shown is adequate to fire from three tanks simultaneously; if it is desired to fire more tanks, targets may be pulled in tandem. The speed of the targets should be approximately 5 miles per hour. The machinegun will be zeroed to hit the center of the target when one lead is taken.

(4) When a moving target range is not available, target tanks may be utilized for this exercise if available and desirable. Ranges for target tanks should be about 400 yards.

Note. The right end box of the range finder is covered, the unit battle sight is set on the range finder, and the machine gun is zeroed at 200 feet. This also applies to the firing of the exercise in (5) and (6) below.

(5) As the targets move along the course, the examining officer gives a fire command, using the same range which was used

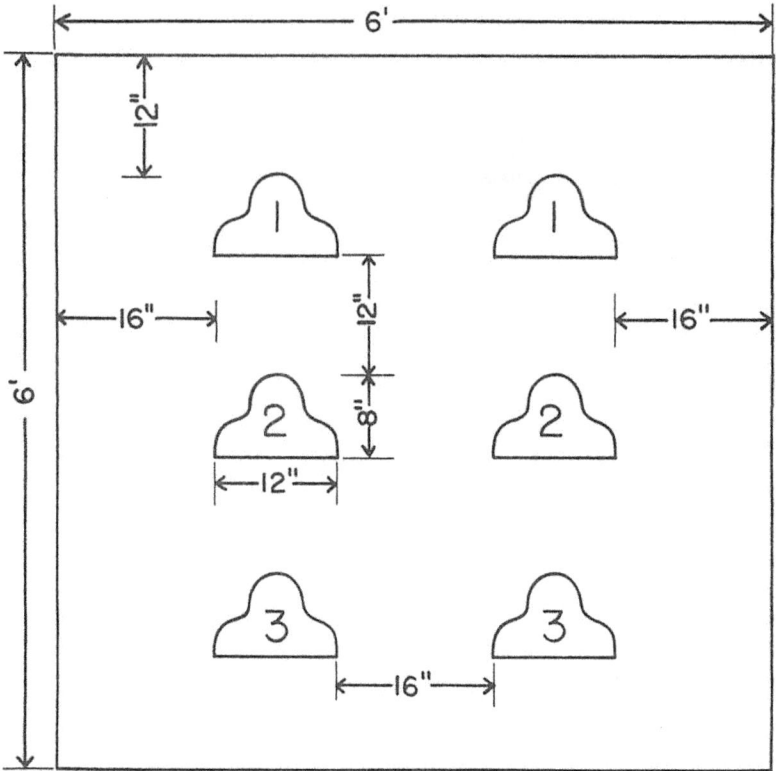

Figure 49. Moving target layout.

to zero the machinegun, and designating the target number of one of the targets. The gunner will then fire three rounds at the target designated. The examining officer records time from the command FIRE until after the third round is fired.

(6) The exercise consists of four trials in (5) above; two trials are fired at the lead target in manual traverse and two at the rear target in power traverse.

(7) The exercise will be scored as follows:

Table II

(Possible score: 100 points)

Trials	Number of rounds	Possible points	Minus	Score
Manual	3	25		
Manual	3	25		
Power	3	25		
Power	3	25		

Table II—Continued

Cuts:

Failure to fire first round within 5 seconds_____ 5 points

Failure to complete trial in 15 seconds_____ 5 points

Each round that fails to hit target_____ 5 points

(8) See figure 50 for illustration of an appropriate score card.

c. *Table III: Moving Tank Exercise (Stationary Target).*

   (1) The purpose of this exercise is to test the gunner's ability to fire the coaxial machinegun from a moving tank at stationary targets.

Co   *A*

Bn   *1 ST*

NAME   *SMITH JOHN J.*

RANK   *PFC*   SN *38132977*

DATE   *2 FEB 54*

90
Total Score

TANK GUNNERY QUALIFICATION COURSE SCORE CARD

100 points
Possible

TABLE II (SUBCALIBER SHOT ADJUSTMENT MOVING TARGET EXERCISE—200 FEET)

TRIALS	NUMBER OF ROUNDS	POS-SIBLE	1ST ROUND FIRED IN 5 SEC		TRIAL COMPLETED IN 15 SEC		TARGET HITS			SCORE
			YES	NO	YES	NO	1ST RD	2D RD	3D RD	
1st—Manual	3	25	X		X		0	5	5	20
2d—Manual	3	25	X		X		5	5	5	25
3d—Power	3	25	X		X		5	5	0	20
4th—Power	3	25	X		X		5	5	5	25
									TOTAL SCORE	90

A. Five points for each "Yes."

B. Five points for each Hit.

Satisfactory Score _____ 70 points.

Lt Leonidas N. James
Examining Officer's Signature

Figure 50. Score card for table II.

(2) In this exercise, the gunner fires 150 rounds from the coaxial machinegun, while the tank is moving, at targets representing infantry.

(3) The length of the course (fig. 51) will be approximately 800 yards. The end of the course will be marked by a white flag on each side of the course. Targets will be kneeling-type (E) silhouettes. Five groups of four silhouettes will be placed not more than 10 yards nor less than 5 yards from the sides of the tank runway, alternately on the left and right sides. One of these groups will be placed at each of the following ranges from the starting point: 200 yards, 350 yards, 450 yards, 550 yards, and 700 yards. A sixth group of five silhouettes will be placed 200 yards beyond the end of the course and in direct line with the center of the course. Each group of silhouettes should cover an area 4 yards wide and 4 yards deep. A red flag will be placed on the edge of the runway, 50 yards from each silhouette group in the direction of the starting line.

(4) The terrain selected for the course will be such that the tank can maintan an average speed of 5 miles per hour.

(5) The ammunition will be loaded four ball to one tracer.

(6) The test is conducted as follows:

(a) The gunner will make only one run over the course while being tested. Fire will cease on a target group when the tank reaches the red flag 50 yards from that group. All hatches will be closed between the starting line and the white flags.

(b) The tank will not stop until it reaches the white flags. At this point firing will cease and the gun will be cleared. The tank will be required to maintain an average speed of 5 miles per hour between the starting line and the white flags.

(c) Separate courses may be set up and used concurrently, provided they are at least 30 yards apart.

(d) The examining officer should follow behind the tank in a vehicle and control its movement by radio. Assistants will follow the tank to mark targets and score.

(e) An assistant will check the number of rounds in each belt before the tank begins the run.

(f) For best results, the coaxial machinegun should be zeroed at 100 yards.

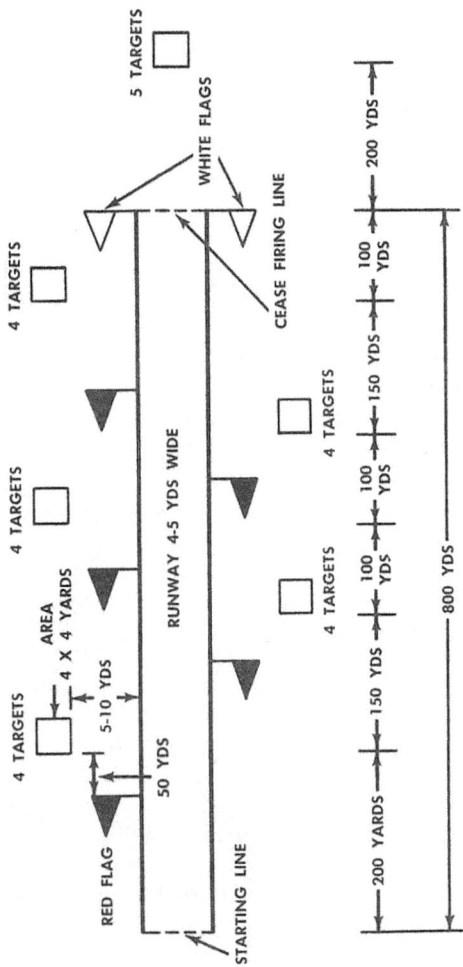

Figure 51. Range setup—moving tank exercise.

(7) Scoring is as follows:

Table III

(Possible score: 100 points)

Target groups	Number of rounds	Possible points	Minus	Score
1	25	16		
2	25	16		
3	25	16		
4	25	16		
5	25	16		
6	25	20		
Total		100		

(a) Four points will be awarded for each silhouette that is hit regardless of the number of hits in each target. The maximum score is 100 points.

(b) No credit will be given for hits on a target group if that target group was fired on after the tank passed the red flag for that group.

(c) The gunner receives no credit for the course if his tank fails to sustain a 5-mile-per-hour speed, but he will be retested.

(8) See figure 52 for illustration of an appropriate score card.

d. *Table IV: Auxiliary Fire Control Exercise.*

(1) *Purpose.* The purpose of this exercise is to test the ability of the gunner in the proper use of the tank's auxiliary fire-control instruments.

(2) *Target layout.*

(a) The impact area selected should be fairly flat, and the surface should be of dirt or sand.

(b) An aiming post, to serve as a gun reference point and 1000-yard range marker, is placed midway within the width of the individual tank's impact area at a distance of 117 feet from the muzzle of the coaxial machine gun.

(c) At varying simulated ranges (500–1500 yards), small vertical targets (4 x 6-inch cards) will be fastened to the ground. The bottom edges of the targets will be folded to provide a fastening surface and a 4 x 4-inch target facing the tank. See figure 53 for target layout.

(3) *Preparation by the examining officer.*

(a) With the coaxial machinegun properly mounted and adjusted, the examining officer fires until a hit is obtained on the base of the gun reference point.

Note. A 4 x 4-inch target is placed immediately behind the aiming post as an aid in zeroing.

Co *A*
Bn *1ST*

NAME *SMITH JOHN J.*
RANK *PFC* SN *38132977*
DATE *3 FEB 54*

80
Total Score

TANK GUNNERY QUALIFICATION COURSE SCORE CARD

100 points
Possible

TABLE III (MOVING TANK EXERCISE, STATIONARY TARGETS)

TARGET GROUPS	NUMBER OF ROUNDS	POSSIBLE	TARGET HITS					SCORE
			NR 1	NR 2	NR 3	NR 4	NR 5	
1	25	16	4	4	0	4		12
2	25	16	4	0	4	4		12
3	25	16	4	4	4	4	(Applies to target group Nr 6 only)	16
4	25	16	0	4	4	4		12
5	25	16	4	4	4	0		12
6	25	20	4	4	4	0	4	16
						TOTAL SCORE		80

A. Four points for each target hit. (No credit for target hits after passing red flag of each group.)

B. No credit if tank does not maintain a speed of 5 mph.

Satisfactory Score _____ 70 points.

Lt Leonidas K. James
Examining Officer's Signature

Figure 52. Score card for table III.

(*b*) Without disturbing the lay of the gun, the examining officer will:

1. Center the bubble of the elevation quadrant with the micrometer knob.

2. Loosen the micrometer knob and slip the scale until a reading of 50 mils appears opposite the index, then tighten the micrometer knob.

334854—55——12

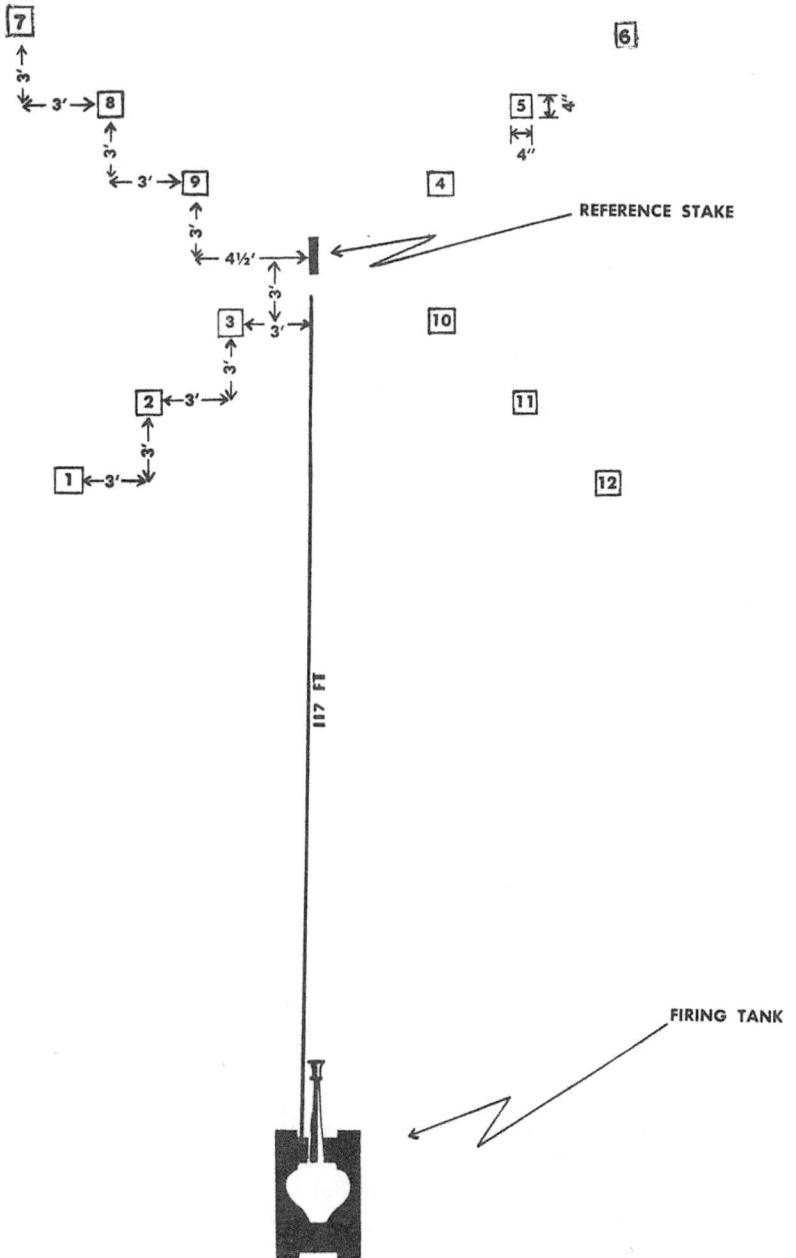

Figure 53. Target layout, table IV.

3. Place a fine chalk or pencil line on the elevation scale opposite the index. This is to avoid 100-mil errors when the gun is elevated between problems.

4. Zero the azimuth indicator.

5. Cover the right end box of the range finder and move the aiming cross, using the boresight knobs, to the point of strike on the gun reference point.

(*c*) After zeroing on the reference point, the examining officer will determine the azimuth indicator reading and elevation quadrant reading to the center of each target in the impact area and will record this data on a reference card (fig. 54). The elevations are converted to ranges, which are also recorded on the reference card.

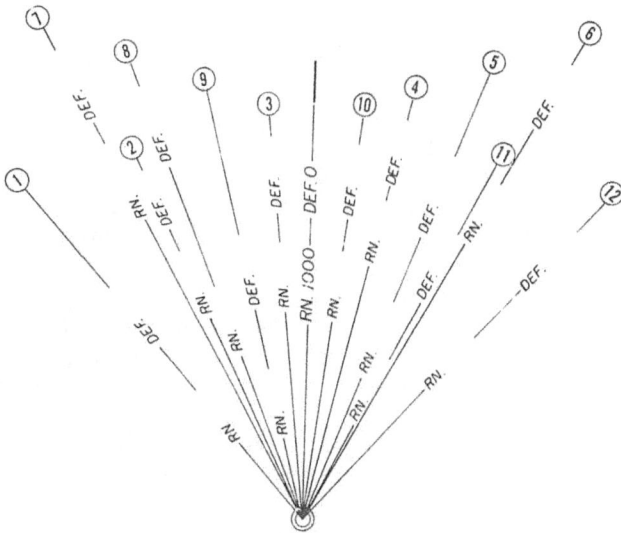

Figure 54. Reference card for table IV.

Note 1. Make certain it is clearly understood this is a REFER-ENCE card and *not* a RANGE card.

Note 2. Elevations are converted to ranges as follows:

(1) Elevation for a range of 1000 yards equals 50 mils.

(2) A 100-yard range change is effected by making a 1-mil change in elevation.

(*d*) The gunner will use a 1-mil change in elevation for each 100 yards of range change desired.

(4) *Procedure for testing.*

(*a*) The gunner will be required to engage four targets, using the auxiliary fire-control instruments. The direct-fire sights will be covered during the exercise. Three rounds of caliber .30 frangible ammunition will be fired at each

167

target; tracer ammunition may be used if caliber .30 frangible ammunition is not available. (The single-shot device can be used, or dummy rounds can be inserted between tracer rounds to permit firing single shots.)

(b) The examining officer, acting as tank commander, will issue correct initial fire commands, sense each round, and give the necessary subsequent commands for each target. For example:

GUNNER

HE

1500 (Correct range to target.)

FROM REFERENCE POINT, RIGHT 30 (This will be based on previously computed data.)

ANTITANK

FIRE. (Time will be recorded from this command.) Subsequent commands will be in accordance with the alternate method of adjustment. Time for subsequent rounds will also be recorded from the command, FIRE.

(c) The gunner, in executing these commands, will use the specially prepared firing table (Note 2, (3) (c) above) and the elevation quadrant for initial subsequent ranges.

(d) In making initial and subsequent deflection shifts, the gunner will use the manual traversing control and the azimuth indicator. At the conclusion of firing on each target, the gun is returned to the reference point and the bubble in the elevation quadrant is centered with a reading of 50 mils.

(5) *Scoring.* Scoring for each problem will be conducted as follows:

Table IV

(Possible score: 100 points)

Trial	Number of rounds	Possible points	Minus	Score
Target 1	3	25		
Target 2	3	25		
Target 3	3	25		
Target 4	3	25		
Total		100		

Cuts:

Failure to get first round off in 20 seconds _____ 5 points
Failure to get second round off in 10 seconds _____ 5 points
Failure to get third round off in 10 seconds _____ 5 points
For each target miss over one _____ 5 points

168

(6) *Score card.* See figure 55 for illustration of an appropriate score card.

Co *A*

Bn *1st*

NAME *SMITH, JOHN J.*

RANK *PFC*   SN *38132977*

DATE *3 FEB 54*

80
_____

Total Score

TANK GUNNERY QUALIFICATION COURSE SCORE CARD

100 points
Possible

TABLE IV (AUXILIARY FIRE CONTROL EXERCISE)

TRIALS	Nr OF RDS	POS-SIBLE	1ST RD IN 20 SEC		2D RD IN 10 SEC		3D RD IN 10 SEC		TARGET HITS			CUTS	SCORE
			YES	NO	YES	NO	YES	NO	1ST RD	2D RD	3D RD		
Target 1	3	25		X	X		X		O	X	X	5	20
Target 2	3	25	X		X		X		O	O	X	5	20
Target 3	3	25	X		X			X	X	O	O	10	15
Target 4	3	25	X		X		X		X	X	O	O	25

TOTAL SCORE | 80 |

A. CUTS: Cut five points for each "NO."

B. Cut five points for each target miss over one (maximum 10 points).

Satisfactory Score _____ 70 points.

Lt Leonidas K. Jones
Examining Officer's Signature

Figure 55. Score card for table IV.

109. Service Firing Exercises

The service ammunition tables are fired only after the gunner has qualified in tables I–IV. Ammunition for both practice and record firing must be of the same lot. Targets will be accurately scored during all record firing; BC scopes or similar viewing instruments may be used for this purpose.

a. Table V: Zeroing Exercise, Service Firing.

(1) This exercise is designed to test the gunner's ability to zero the range finder and M20 periscope, using shot ammunition

on a stationary panel target at a known range. The gun and sights must be properly zeroed to obtain maximum effectiveness.

(2) (a) The gunner is required to boresight the range finder and M20 periscope on the zeroing target, which should be at a range as near 1500 yards as possible.

(b) After boresighting and applying the emergency zero setting, the gunner will lay the aiming cross on the aiming point of the zeroing target.

(3) The target is 6 feet square and is designed to provide a well-defined aiming point at a known range (fig. 56).

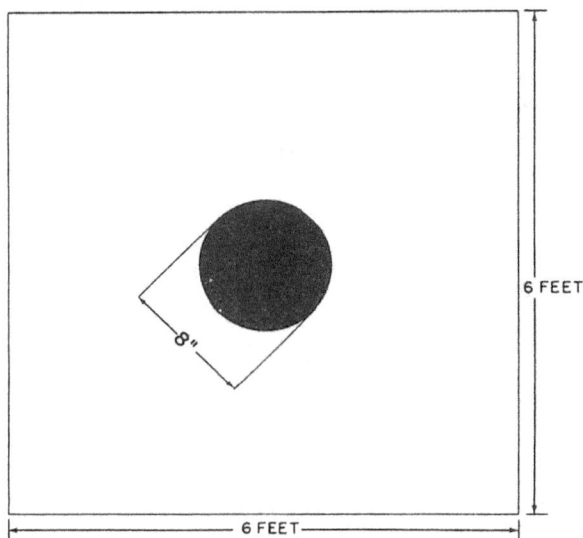

Figure 56. Target panel for zeroing exercise.

(4) (a) The gunner indexes on the range finder the code number for the ammunition and the correct range to the target. He also indexes the correct range on the M3 ballistic unit.

(b) The gunner then lays the aiming cross of the gun laying reticle on the aiming point (center of 8-inch bull's-eye) and fires three rounds to form a shot group, checking the lay after each round.

(c) Without disturbing the lay of the gun, he unlocks the elevation and azimuth boresight knobs on the Range Finder, M12, and Periscope, M20. He moves the aiming crosses of the gun laying reticle to the center of the shot group by manipulating the boresight knobs, and then relocks the boresight knobs.

170

(*d*) The gunner re-lays on the aiming point and fires one check round to determine whether he has zeroed correctly.

(5) There is no time limit for this exercise. Scoring will be accomplished in accordance with table V (M47).

Table V (M47)

(Possible score: 100 points)

No.	Item	Cuts (points)	Possible points	Minus	Score
1	Correct boresight procedure, Range Finder, M12, and Periscopes, M20, total value____	_____	30	_____	_____
	For improper selection of boresight point, cut____	6	_____	_____	_____
	For failure to set range scale of M12 range finder on "B" prior to boresighting, cut_	6	_____	_____	_____
	For inaccuracy in laying aiming crosses of gun laying reticles on boresighting point observed through bore, cut____	6	_____	_____	_____
	For failure to lock boresighting locking levers after placing aiming crosses on boresighting point, cut____	6	_____	_____	_____
	For failure to set the slip scales properly on the boresighting knobs of the range finder and M20 periscope, cut____	6	_____	_____	_____
2	Correct zeroing procedure, Range Finder, M12, and Periscopes, M20, total value____	_____	30	_____	_____
	For failure to index proper emergency zero, cut____	6	_____	_____	_____
	For failure to index proper ammunition code number on ammunition scale of range finder, cut____	6	_____	_____	_____
	For failure to index proper range on range scale of range finder, cut____	6	_____	_____	_____
	For failure to use boresighting knobs in manipulating aiming cross of range finder gun laying reticle on center of shot group, cut____	6	_____	_____	_____
	For failure to index correct range opposite proper type ammunition on ballistic unit, cut____	6	_____	_____	_____
3	Accuracy of zero, total value____	_____	40	_____	_____
	If check round strikes within 14 inches from center of aiming point____	no cut	_____	_____	_____
	More than 14 inches and less than 18 inches from center of aiming point, cut__	10	_____	_____	_____
	More than 18 and less than 24 inches from center of aiming point, cut____	20	_____	_____	_____
	More than 24 inches from center of aiming point, cut____	40	_____	_____	_____
	Total____	_____	100	_____	_____

Note. A physical check of the target will be made to insure positive scoring of No. 3.

(6) See figure 57 for illustration of an appropriate score card.

b. Table VI: Service Firing Exercise, Stationary Targets at Variable Ranges (Shot and HE Adjustment).

(1) This exercise is designed to test the gunner's ability to utilize the range finder and the burst-on-target method of adjustment while firing service ammunition at stationary targets. The exercise is more realistic if tanks move between problems. The gunner will fire at four separate targets (two SHOT and two HE), using the range finder to determine the initial range and, if necessary, the burst-on-target method of adjustment for subsequent rounds. The gunner will be

Co *A*
Bn *1ST*

NAME *SMITH JOHN J.*
RANK *PFC* SN *38132977*
DATE *4 FEB 54*

94
Total Score

TANK GUNNERY QUALIFICATION COURSE SCORE CARD

100 points
Possible

TABLE V (ZEROING EXERCISE FOR M47 TANK)

NR 1	BORESIGHTING	YES	NO	SCORE
(30 points)	1. Properly selected boresight point.	X		6
6 points for each YES.	2. Removed all superelevation prior to boresighting.	X		6
	3. Properly laid aiming cross on boresight point.	X		6
	4. Boresight locking levers locked after placing aiming cross on boresight point.	X		6
	5. Slip scales set properly on boresight knobs of range finder and periscope.	X		6
NR 2	ZEROING PROCEDURE			
(30 points)	1. Indexed proper emergency zero.	X		6
6 points for each YES.	2. Indexed proper ammunition on ammunition scale.		X	0
	3. Indexed proper range on range scale.	X		6
	4. Placed proper range on telescope or periscope.	X		6
	5. Used boresight knobs in manipulating aiming point of reticles on center of shot group.	X		6
NR 3	ACCURACY OF ZERO			
40 points for YES.	Check round within 14 inches from the center of aiming point.	X		40
		TOTAL SCORE		94

Satisfactory Score --- 70 points.

Lt Leonidas H. James
Examining Officer's Signature

Figure 57. Score card for Table V.

172

limited to two rounds of ammunition for each of the four targets he is to engage.

(2) The examining officer will indicate each of the targets by issuing an initial fire command. The examining officer will lay the gun for direction, using the commander's power control handle and the M20 periscope. Time for each problem starts when the command FIRE is announced in the initial fire command.

(3) If the first round is not a target hit, the gunner will use the burst-on-target method of adjustment to fire the subsequent round. If the target is hit on the first round, full credit will be given and the second round *will not be fired.*

(4) Targets to be used for this exercise are 3 x 5-foot cloth panels for taget 1 and 2, and 6 x 6-foot cloth panels for target 3 and 4. One HE and one SHOT problem will be fired at each size target.

(5) Scoring for this exercise will be conducted as indicated in Table VI.

Table VI

(Possible score : 100 points)

Trial	Range (yards)	Number of rounds	Possible points	Minus	Score
Target 1_____	800–1100	2	25	_____	_____
Target 2_____	1100–1500	2	25	_____	_____
Target 3_____	1500–1800	2	25	_____	_____
Target 4_____	1800–2000	2	25	_____	_____
Totals_____	_____	8	100	_____	_____

Cuts:

Failure to fire first round within 15 seconds_____ 5 points

Deduct one point for each additional second over 15 required to fire first round, up to 20 seconds_____ 5 points

Failure to hit target with first round_____ 10 points

Failure to hit target with second round, if fired_____ 5 points

(6) See figure 58 for illustration of an appropriate score card.

c. *Table VII: Moving Target Exercise (SHOT).*

(1) The purpose of this exercise is to test the ability of the gunner to deliver effective fire on a moving target. The gunner fires at five moving targets, using the range finder to determine the initial ranges and the burst-on-target method of adjustment for subsequent rounds. The gunner will be limited to two rounds of ammunition for each of the five targets he is to engage.

Co *A*

Bn *1ST*

NAME *SMITH, JOHN J.*
RANK *PFC* SN *38132977*
DATE *4 FEB 54*

POSSIBLE SCORE   100
TOTAL CUTS   *28*
TOTAL SCORE   *72*

TANK GUNNERY QUALIFICATION COURSE SCORE CARD

TABLE VI (SERVICE FIRING, HE AND SHOT ADJUSTMENT).

| | | | A | B | | C | |
| | NUMBER OF ROUNDS | POSSIBLE POINTS | TIME 1ST ROUND FIRED (Seconds) | TARGET HIT 1ST ROUND | | TARGET HIT 2D ROUND | |
TRIAL				YES	NO	YES	NO
TARGET 1	2	25	*12*	X			
2	2	25	*18*	X			
3	2	25	*11*		X	X	
4	2	25	*9*		X	X	

Cuts:

| | MAXIMUM CUTS | TARGETS | | | | TOTAL CUTS |
		1	2	3	4	
A. Failure to fire 1st round within 15 seconds 5 points		0	5	0	0	5
(Deduct one point for each second over 15.) Maximum cut. 5 points		0	3	0	0	3
B. Failure to hit target with 1st round. 10 points		0	0	10	10	20
C. Failure to hit target with 2d round. 5 points		0	0	0	0	0
(In case the target is hit by 1st round the 2d round will not be fired.)					TOTAL	28

Satisfactory Score .... 70 points.

Lt Leonidas K. James
Examining Officer's Signature

Figure 58. Score card for Table VI.

(2) The exercise is fired from a stationary tank at moving targets (6 x 6-foot panels) at unknown ranges, varying from 700 to 1500 yards. Either a powered target or a towed target may be used. To vary the range from tank to target, either the tank or the targets may be moved to different locations. The target will be exposed for approximately 300 yards and will travel that distance at a constant speed of between 8 and 15 miles per hour.

(3) The examining officer will lay the gun for direction for each target while issuing a five-element initial fire com-

mand. Time for each problem starts when the command
FIRE is announced in the initial fire command.
   (4) The gunner will use the burst-on-target method of adjust-
      ment to fire the subsequent round. This round will be fired
      even if the first round is a target hit.
   (5) Scoring will be conducted as indicated in Table VII.

Table VII

(Possible score: 100 points)

Trial	Number of rounds	Possible points	Minus	Score
Target 1_____	2	20	_____	_____
Target 2_____	2	20	_____	_____
Target 3_____	2	20	_____	_____
Target 4_____	2	20	_____	_____
Target 5_____	2	20	_____	_____

Cuts:
   Failure to fire first round within 15 seconds_____ 5 points
   Deduct 1 point for each additional second over 15 to fire first round,
      up to 20 seconds_____ 5 points
   Failure to hit target with first round_____ 5 points
   Failure to hit target with second round_____ 5 points

   (6) See figure 59 for illustration of an appropriate score card.
d. Table VIII: Range Card Firing Exercise.
   (1) The purpose of this exercise is to test the ability of the gunner
      to determine prearranged firing data to selected targets and
      to engage area type targets successfully with HE ammunition
      under conditions of restricted visibility, and to afford night
      firing practice.
   (2) Five 6 x 6-foot panels will be placed in a wide lateral area,
      at ranges varying from 800 yards minimum to 3,500 yards
      maximum, and at different angles of site. Panels will be
      numbered consecutively from left to right and will be visible
      from a suitable firing position.
   (3) Examining personnel will accurately compute the following
      data for each panel.
      (*a*) The azimuth indicator reading from an aiming stake or
         reference point.
      (*b*) The quadrant elevation, gun to target (elevation for range
         plus angle of site), with the elevation quadrant. This is
         determined in the following manner:
         1. Check and adjust the elevation quadrant with the gunner's
            quadrant.
         2. Determine the range to the target with the rangefinder.

NAME SMITH, JOHN J.
RANK PFC   SN 38132977
DATE 5 FEB 54

POSSIBLE SCORE	100
TOTAL CUTS	17
TOTAL SCORE	83

TANK GUNNERY QUALIFICATION COURSE SCORE CARD

TABLE VII (SERVICE FIRING MOVING TARGET).

TRIAL	NUMBER OF ROUNDS	POSSIBLE POINTS	TIME 1ST ROUND FIRED (Seconds)	A — TARGET HIT 1ST ROUND — YES	A — NO	B — TARGET HIT 2D ROUND — YES	B — NO
TARGET 1	2	20	11		X	X	
2	2	20	12	X		X	
3	2	20	14	X		X	
4	2	20	17	X		X	
5	2	20	13		X	X	

Cuts:

	MAXIMUM CUTS	TARGETS 1	2	3	4	5	TOTAL CUTS
A. Failure to fire 1st round within 15 seconds	5 points	0	0	0	5	0	5
(Deduct one point for each second over 15.) Maximum cut.	5 points	0	0	0	2	0	2
B. Failure to hit target with 1st round	5 points	5	0	0	0	5	10
C. Failure to hit target with 2d round	5 points	0	0	0	0	0	0

TOTAL 17

Satisfactory Score .... 70 points.

Lt Leonidas K. Jones
Examining Officer's Signature

Figure 59. Score card for Table VII.

3. Lay the aiming point of the sight reticle on the center of the target, as would be done in direct fire.

4. Without disturbing the lay of the gun, measure the existing elevation with the elevation quadrant.

> *Note.* If the reading on the micrometer dial is between markings, record to the next higher whole mil.

(4) Ten E-type silhouette targets will be placed around the panel at which the gunner will fire. These panels will be placed as shown in figure 60.

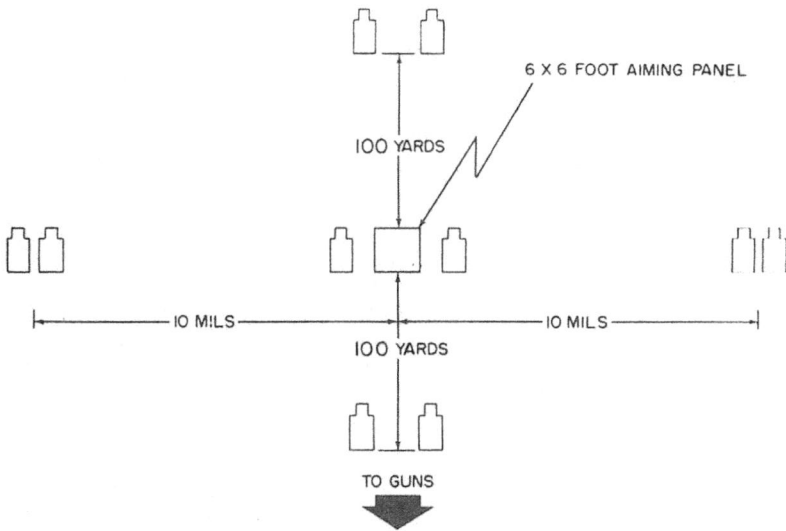

Figure 60. Range card firing exercise.

(5) The exercise will be conducted as follows:

  (a) *Part I.* The gunner will be required to prepare a card for the area, using the panels as likely targets. Information to be recorded on the range card will include—

  1. Aiming stake or reference point.

  2. Target (panel) number (left to right).

  3. Deflection (azimuth indicator reading) from aiming stake.

  4. Range to target in yards.

  5. Quadrant for HE ammunition.

  (b) *Part II.* After the range card is prepared, the direct-fire sights will be covered (or the exercise may be fired at night), and the gunner will be required to fire on one of the panels, using his prearranged firing data. The examining officer will issue an initial fire command, using the data computed by the gunner, to one of the targets (see (4) above). The gunner will set off the data as announced in the initial fire command and fire only the first round. The gunner will then add one mil in elevation and simulate firing the second round; drop two mils and simulate firing the third round; add one mil, traverse right 10 mils, and simulate firing the fourth round; traverse left 20 mils and simulate firing the fifth round. He will announce ON THE WAY as he simulates firing each round. Time will be recorded from the command FIRE.

  Note. Elevation will be changed 1 mil to effect the 100-yard range change.

(6) Scoring will be conducted as follows:

Table VIII

(Possible score: 100 points)

No.	Item	Possible cuts	Possible points	Minus	Score
Part I__	Total value_____	_____	50	_____	_____
	For failure to obtain correct azimuth indicator reading within plus or minus 1 mil on each target, cut 5 points for each mil (maximum cut 25 points).	25	_____	_____	_____
	For failure to obtain correct quadrant reading within plus or minus 1 mil on each target, cut 5 points for each mil (maximum cut 25 points).	25	_____	_____	_____
Part II__	Total value_____	_____	50	_____	_____
	For each five seconds, or fraction thereof, that the time to fire the first round exceeds 25 seconds, cut 5 points (maximum cut 25 points).	25	_____	_____	_____
	For failure to hit in target area (20 mils by 200 yards), cut.	25	_____	_____	_____

Possible_____ 100 points

Cuts_____ _____

Score_____

(7) See figure 61 for illustration of an appropriate score card.

Co *A*
Bn *13T*

NAME *SMITH JOHN J.*
RANK *PFC* SN *38132977*
DATE *6 FEB 54*

POSSIBLE SCORE   100
TOTAL CUTS   *15*
TOTAL SCORE   *85*

TANK GUNNERY QUALIFICATION COURSE SCORE CARD

TABLE VIII (RANGE CARD FIRING EXERCISE)

PART I		ROUNDS					TOTAL CUTS
		1	2	3	4	5	
(50 points)	1. Failure to obtain correct azimuth indicator reading within plus or minus 1 mil on each target. (Cut 5 points for each mil error not to exceed a total of 25 points.)	5	0	0	5	0	*10*
	2. Failure to obtain correct quadrant reading within plus or minus 1 mil on each target. (Cut 5 points for each mil error not to exceed a total of 25 points.)	0	5	0	0	0	*5*
PART II		TIME FIRST RD FIRED (SEC)					
(50 points)	1. Failure to fire first round within 25 seconds. (Cut 5 points for each five seconds or fraction thereof over 25 seconds not to exceed 25 points.)	*19*					*0*
	2. Failure to hit target area 20 mils by 200 yards. (Cut 25 points.)	*HIT*					*0*
						TOTAL CUTS	*15*

Satisfactory Score .... 70 points.

Lt Leonidas K. James
Examining Officer's Signature

Figure 61. Score card for Table VIII.

APPENDIX I

REFERENCES

DA Pam 108–1	Index of Army Motion Pictures, Television Recordings and Filmstrips.
DA Pam 310–1	Index of Administrative Publications.
DA Pam 310–3	Index of Training Publications.
DA Pam 310–4	Index of Technical Manuals, Technical Regulations, Technical Bulletins, Supply Bulletins, Lubrication Orders, and Modification Work Orders.
SR 310–20–6	Index of Blank Forms.
SR 320–5–1	Dictionary of United States Army Terms.
SR 320–50–1	Authorized Abbreviations.
SR 385–310–1	Regulations for Firing Ammunition for Training, Target Practice, and Combat.
FM 5–25	Explosives and Demolitions.
FM 17–12	Tank Gunnery.
FM 21–5	Military Training.
FM 21–6	Techniques of Military Instruction.
FM 21–8	Military Training Aids.
FM 21–30	Military Symbols.
FM 21–60	Visual Signals.
FM 23–55	Browning Machineguns, Caliber .30, M1917A1, M1919A4, and M1919A6.
FM 23–65	Browning Machineguns, Caliber .50, HB, M2.
TM 9–718A	90-mm Gun, Tank, M47.
TM 9–1901	Artillery Ammunition.
TM 11–284	Radio Sets AN/GRC–3, –4, –5, –6, –7, and –8.
TM 11–704	Auxiliary Interphone Equipment, AN/VIA–1.

APPENDIX II

SUBJECT SCHEDULE

1. General

The following subject schedule is presented for use by the unit commander, operations officer, or individual instructor. It is a guide in training the individual crewman and the tank crew in all phases of operation of the Tank, 90-mm Gun, M47. The schedule can easily be expanded if more instructional time is available. With the schedule followed as written, the maximum operating efficiency will be attained by the tank crews. More time should be used on the performance of crew duties if thorough proficiency is the training objective.

2. Tank Crew and Individual Crewman Training, Excluding Driver Training

Unless otherwise specified, all references are to this field manual. Figures in parentheses under *Hours* indicate concurrent training.

Period	Hours	Subject	References	Area	Training aids and equipment
1	4	Turret familiarization	Chap. 2, sec. I, III, IV, V	Tank park	One tank per five students.
2	4	Disassembly and assembly of 90-mm Gun, M36.	Chap. 2, sec. II	Tank park	One tank per five students.
3	2	Functioning and malfunctions of 90-mm Gun, M36.	Chap. 2, sec. II	Classroom	One set charts of 90-mm Gun, M36.
4	3	Hydraulic traversing and elevating mechanism.	Chap. 2, sec. VI	Tank park, classroom	One tank per five students; one set of schematic drawings of hydraulic mechanism.
5	4	Direct-fire control equipment.	Chap. 2, sec. V	Tank park, classroom	One tank per five students; one set of charts on boresighting and zeroing.
6	3	Auxiliary fire control equipment.	Chap. 2, sec. V	Tank park	One tank per five students.
7	56 (16)	Range Finder, M12	Chap. 2, sec. V	Classroom, range determination area.	One set range finder charts; one tank per five students; range finder trainer.
8	1	Crew composition, formations, and control.	Chap. 3, sec. I, II	Classroom	Formation charts and control charts.
9	2 (6)	Crew drill	Chap. 3, sec. III	Classroom, tank park, or training area.	Charts of duties for each crew member; one tank per five students.
10	3 (6)	Mounted action	Chap. 3, sec. IV, V	Classroom, tank park, or training area.	Charts of duties for each crew member; one tank per five students.
11	1	Dismounted action	Chap. 3, sec. VI	Classroom, tank park, or training area.	Charts of duties for each crew member; one tank per five students.

12	1	Evacuation of wounded	Chap. 3, sec. VII	Classroom, tank park, or training area.	Charts of duties for each crew member; one tank per five students.
13	1	Destruction of equipment	Chap. 3, sec. IX	Tank park or training area.	One tank.
14	4 (4)	Stowage (OVM) (ammunition).	Chap. 3, sec. X	Tank park or training area.	One tank per five students.
15	Unit training SOP	Inspections and maintenance.	Chap. 3, sec. VIII	Tank park or training area.	Charts of duties for each crew member; one tank per five students.
16	8 (6)	Conduct of fire: *a.* HE adjustment, shot adjustment, and firing of machineguns. *b.* Moving targets (burst-on-target and alternate methods).	Chap. 4, sec. I, II, III; FM 17–12.	Classroom or training area.	Terrain board with equipment; burst-on-target trainer; one tank or turret trainer per five students.
17	4 (4)	Crew nonfiring exercises	Chap. 4, sec. V	Tank park or training area.	One tank or trainer per five students.
18	16	Gunner's preliminary examination.	Chap. 4, sec. V	Tank park or training area.	One tank or trainer per five students.
19	8	Subcaliber firing, manipulation, and shot adjustment at stationary targets.	Chap. 4, sec. V	Range	One tank per five students; subcaliber range equipment and ammunition; see references.
20	6	Service firing zeroing exercises; HE and shot adjustment.	Chap. 4, sec. I, II, III, V	Range	One tank per five students; normal range equipment and ammunition; see references, pertinent range regulations.

[1] Periods 9 and 10 should be scheduled in conjunction with all periods of service firing.

Period	Hours	Subject	References	Area	Training aids and equipment
21	6	Service firing at moving targets.	Chap. 4, sec. I, II, IV, V ----	Range-----------	See period 20.
22	6	Service firing------------	Chap. 4, sec. I, II, III, V ----	Range-----------	See period 20.

Total hours of instruction: 143.
Total hours of concurrent training: (42).
Concurrent training on periods 7, 9, 10, 14, 16, and 17 to be scheduled during range firing.

APPENDIX III

NOMENCLATURE OF COMPONENTS

Current technical manuals pertaining to the M47 tank list certain components by their T-numbers; in some instances, the nomenclature plate on the component will bear the T-number. Since publication of these manuals, the components have been standardized. For reference purposes, the T-numbers of components are given below opposite their M-number counterparts.

Gun, 90-mm, M36	Gun, 90-mm, T119E1
Range Finder, M12	Range Finder, T41
Periscope, M20	Periscope, T35
Elevation Quadrant, M13	Elevation Quadrant, T21
Azimuth Indicator, M31	Azimuth Indicator, T24
Ballistic Drive, M3	Ballistic Drive, T23E1
Superelevation Transmitter, M22	Superelevation Transmitter, T13
Mount, Periscope, M88	Mount, Periscope, T176
Mount, Periscope, M89	Mount, Periscope, T177

INDEX

	Paragraph	Page
Abandon tank (drill)	75	109
Accumulator:		
Hand pump	21	29
System, elevation mechanism	40	57
Action:		
Dismounted (drill)	73	104
Mounted (drill)	67	96
Additional inspection and maintenance	86	121
Adjustment:		
Coaxial machinegun firing solenoid	13	19
Elevation quadrant	36	51
Examination	104	151
Fire:		
Alternate method	96	142
Moving target	94	138
Primary method	95	139
Gunner's quadrant	37	55
Examination	104	151
Periscope	30	40
Range finder	29	32
Sights, examination	104	151
Synchronization and backlash	34	50
After-firing checks, 90-mm gun	11	18
After-operation inspection	85	120
Alternate method, adjustment of fire	96	142
Ammunition:		
Examination	104	151
Scale:		
Periscope	30	40
Range finder	29	32
Tank gunnery qualification course	101	150
Use, handling, stowage	72	103
Apparent speed	94	138
Armament controls. (*See* Controls.)		
Assembly:		
Breechblock	104	151
Flareless tube fittings	42	82
Gun, 90-mm	7	8
At-the-halt inspection	84	117
Auxiliary fire control equipment	28, 36	32, 51
Exercise	107	156
Azimuth:		
Boresight knob, range finder	29	32
Indicator	36	51
Examination	105	153
Backlash, linkage	34	50

	Paragraph	Page
Ballistic:		
Correction knob, range finder	29	32
Drive	30	40
Examination	105	153
Battle sights, range finder	29, 91	32, 134
Before-firing checks, 90-mm gun	11	18
Before-operation inspection	82	111
Bleeding:		
Manual elevation system	40	57
Recoil system, 90-mm gun	9	14
Boresighting	31	45
Coaxial machinegun	13	19
Machineguns	107	156
Bow gunner. (*See also* Crew members.)		
Firing bow gun	97	146
Bow machinegun. (*See* Machinegun.)		
Box, power control	23	29
Breech:		
Functioning	8	13
Opening	62	94
Breechblock, disassembly and assembly	104	151
Burst-on-target method of fire	95	139
Caliber .50 machinegun. (*See* Machinegun.)		
Care, 90-mm gun	11	18
Care-and-maintenance examination	104	151
Characteristics, M47 tank	2	3
Checking:		
Interphone equipment	50	85
Radio	49	85
Recoil system, 90-mm gun	9	14
Classification of gunners	100	149
Cleaning 90-mm gun	11	18
Close hatches (drill)	55	89
Closing of breech, 90-mm gun	8	13
Coaxial machinegun. (*See* Machinegun.)		
Cocking, 90-mm gun	8	13
Combination gun mount	13	19
Commands:		
Control of turret	52	85
Fire. (*See* Fire commands.)		
Movement of tank	52	85
Commander's periscope. (*See* Periscope.)		
Components:		
Combination gun mount	13	19
Nomenclature	app III	185
Composition of crew	44	83
Conduct of fire	89	133
Control(s):		
Armament	18, 38	26, 57
Box:		
Interphone	47, 48	84
Power	23	29
Firing	24	30

	Paragraph	Page
Control(s)—Continued		
Gunner's:		
Manual	21, 38, 40	29, 57
Power	19, 38, 41	27, 57, 66
Power, data	39	57
Range finder	29	32
Tank commander's	20, 38, 41	28, 57, 66
Turret	18, 38	26, 57
Correction of malfunctions, 90-mm gun	10	17
Course, tank gunnery qualification	99	148
Crew:		
Composition	41	66
Drill	43	83
Dismounted action	73	104
Evacuation of wounded	79	110
Mounted action	67–72	96
Formations	45	83
Gun, mounted positions	60	93
Members, duties:		
During inspection and maintenance	80–86	110
Firing	69, 90, 91, 97	99, 133, 134, 146
Record firing	102	150
Subject schedule for training	app II	181
Data:		
Gun, 90-mm	6	8
Power controls	39	57
Range finder	29	32
Tank, M47	4	4
Defilade, firing from	98	148
Description:		
Azimuth indicator	36	51
Elevation quadrant	36	51
Gun, 90-mm	5	8
Gunner's quadrant	37	55
Tank, M47	3	4
Destroy tank (drill)	76	109
Destruction of equipment	87	123
Diopter adjustment:		
Periscope	30	40
Range finder	29	32
Direct fire. (*See* Fire.)		
Sights	28	32
Disassembly:		
Breechblock	104	151
Gun, 90-mm	7	8
Dismounted:		
Action (drill)	73	104
Crew drill	53	86
Posts, crew	45	83
Dismounting crew (drill)	56, 57	91, 92
Dither pump	41	66
Doubtful sensing	92	137
Drill, crew. (*See* Crew Drill.)		

	Paragraph	Page
Driver (*see also* Crew members):		
Commands to	52	85
During-firing checks, 90-mm gun	11	18
During-operation maintenance	83	117
Duties, crew members. (*See* Crew members.)		
Elements, fire commands:		
Initial	91	134
Subsequent	93	137
Elevating controls. (*See* Controls.)		
Elevation:		
Boresight knob, range finder	29	32
No-back	40	57
Quadrant	36	51
Examination	104, 105	151, 153
End boxes, range finder	29	32
End-for-end test	37	55
Equipment:		
Destruction	87	123
Stowage	88	123
Escape hatches (drill)	57	92
Evacuation of wounded from tank	77	109
Drill	79	110
Examination:		
Gunner's preliminary	103	151
Scoring	106	156
Materiel	104	151
Simulated firing	105	153
Exercises, firing	107, 108	156, 157
Extraction and ejection, 90-mm gun	8	13
Failures to function, 90-mm gun	10	17
Fight on foot (drill)	73	104
Filling recoil system, 90-mm gun	9	14
Filter lever, range finder	29	32
Fire:		
Adjustment of. (*See* Adjustment.)		
Commands	89	133
Initial	91, 94	134, 138
Machinegun fire	97	146
Subsequent	93, 94	137, 138
Conduct of	89	133
Control equipment	28, 36	32, 51
Special conditions	98	148
Firing:		
Coaxial machinegun	13	19
Controls. (*See* Controls.)		
Duties of crew in	69, 90	99, 133
Exercises	107, 108	156, 157
Gun, 90-mm	8	13
Solenoid, coaxial machinegun	13	19
Flareless tube fittings	42	82
Formations, crew	45	83
Forms, maintenance	81	111

	Paragraph	Page
Functioning:		
Gun, 90-mm	8	13
Manual controls	40	57
Power controls	41	66
Gun:		
Crew. (*See* Crew.)		
Laying reticle	29	32
Machine. (*See* Machinegun.)		
Mount. (*See* Mount.)		
90-mm:		
Assembly	7	8
Data	6	8
Description	5	8
Disassembly	7	8
Functioning	8	13
Loading	63	94
Switch	23	29
Traveling lock	26	31
Gunner('s). (*See also* Crew members.)		
Classification of	100	149
Commands to	52	85
Controls. (*See* Controls.)		
Periscope. (*See* Periscope.)		
Preliminary examination	103	151
Quadrant	36, 37	51, 55
Examination	104, 105	151, 153
Range finder training	33	47
Safety precautions	61	93
Gunnery qualification course, tank	99	148
Halving knob, range finder	29	32
Handles, control. (*See* Controls.)		
Handling ammunition	72	103
Hatches:		
Dismount through escape (drill)	57	92
Open and close (drill)	55	89
HE adjustment exercise	108	157
History, M47 tank	2	3
Importance, proper stowage	88	123
Index, range finder	29	32
Indicator, oil, 90-mm gun	9	14
Initial fire commands	91	134
Moving target	94	138
Inspections	80–86	110
Examination	104	151
Installation:		
Bow machinegun	15	26
Caliber .50 machinegun	14	24
Coaxial machinegun	13	19
Range finder end boxes	29	32
Instrument light:		
M36	30	40
T22	36	51
Internal correction system, range finder	29	32

	Paragraph	Page
Interphone:		
Control box	47, 48	84
Equipment, checking	50	85
Language	52	85
Operation	46, 48	84
Interpupillary adjustment, range finder	29	32
Inverter	29, 41	32, 66
Lamps, reticle, range finder	29	32
Language, interphone	52	85
Leading	94	138
Lever, hand firing	24	30
Light switch, range finder	29	32
List, stowage	88	123
Load all weapons (drill)	71	103
Loader('s) (see also Crew members):		
Reset safety	24	30
Safety precautions	61	93
Traverse safety	22	29
Loading:		
Coaxial machinegun	13	19
Gun, 90-mm	63	94
Locks	26	31
Lost sensing	92	137
Lubrication	80–86	110
Examination	104	151
Gun, 90-mm	11	18
Machinegun(s)	12	19
Boresighting	107	156
Bow	15	26
Caliber .50	14	24
Coaxial	13	19
Switch	23	29
Firing	97	146
Subcaliber firing	107	156
Zeroing	107	156
Maintenance	80–86	110
Examination	104	151
Gun mounts	17	26
Malfunctions, 90-mm gun	10	17
Manipulation exercise	107	156
Manual controls. (See Controls.)		
Master relay switch	25	31
Materiel examination	104	151
Mechanism, recoil, 90-mm gun	9	14
Method. (See Procedure.)		
Misfire, unloading	64	95
Modes of operation, interphone control box	48	84
Monitoring interphone and radio	48	84
Motor, turret electric	41	66
Mounted:		
Action	67	96
Posts, crew	45	83
Mounting tank crew	54	88

	Paragraph	Page
Mounts:		
Combination gun	13	19
Gun, maintenance	17	26
Machinegun	12	19
Tripod	16	26
Periscope	30	40
Pintle, caliber .50 machinegun	14	24
Movement of tank, commands for	52	85
Moving target:		
Exercise	107, 108	156, 157
Firing at	94	138
Subsequent fire command	93	137
Night firing	98	148
No-back mechanism	21, 40	29, 57
Nomenclature of components	app III	185
Oil indicator, 90-mm gun	9	14
Open hatches (drill)	55	89
Opening of breech, 90-mm gun	8	13
Procedure	62	94
Operation (see also Procedure):		
Interphone and radio	46, 48	84
Range finder	29	32
Over sensing	92	137
Override lever	20, 41	28, 66
Parts, stowage	88	123
Pep drill	58	93
Periodic inspection and maintenance	86	121
Periscope:		
Boresighting	31	45
M6–M13	35	51
M20	30	40
Examination	105	153
Zeroing	32	46
Pintle mount, caliber .50 machine gun	14	24
Positions:		
Gun crew	60	93
Interphone control box	47	84
Possible scores, gunner's preliminary examination	106	156
Post, crew drill	45	83
Power:		
Controls. (See Controls.)		
Operation, turret	27	31
Examination	104	151
Precautions, safety	61	93
Preliminary examination, gunner's	103	151
Premature firing, 90-mm gun	10	17
Prepare to fire (drill)	68	96
Preparation for mounted action (drill)	67	96
Primary method, adjustment of fire	95	139
Principles of functioning. (See Functioning.)		

	Paragraph	Page
Procedure:		
Adjusting fire:		
Alternate method	96	142
Primary method	95	139
Boresighting	31, 107	45, 156
Destruction of equipment	87	123
Disassembly and assembly, 90-mm gun	7	8
Evacuating wounded from tank	78	110
Installing end boxes, range finder	29	32
Loading 90-mm gun	63	94
Opening breech	62	94
Operation of range finder	29	32
Power elevation and traverse	19	27
Range finder training	33	47
Removing stuck:		
Projectile	65	95
Round	66	95
Turret power operation	27	31
Unloading unfired round or misfire	64	95
Zeroing	32, 107	46, 156
Projectile, stuck, removing	65	95
Pumps(s):		
Accumulator hand	21	29
Power control system	41	66
Purpose of:		
Manual	1	3
Tank gunnery qualification course	99	148
Putting turret into power operation	27	31
Quadrant:		
Elevation	36	51
Examination	104, 105	151, 153
Gunner's	36, 37	51, 55
Examination	104, 105	151, 153
Qualification course, tank gunnery	99	148
Radio:		
Check	49	85
Operation	46, 48	84
Range:		
Card firing exercise	108	157
Finder	29	32
Boresighting	31	45
Examination	105	153
Training	33	47
Zeroing	32	46
Knob, rangefinder	29	32
Recoil mechanism, 90-mm gun	9	14
Record firing	100	149
Rules for	102	150
References	app I, III	180, 185
Relay switch, master	25	31
Remount (drill)	74	107

	Paragraph	Page
Removal:		
Bow machinegun	15	26
Caliber .50 machinegun	14	24
Coaxial machinegun	13	19
Projectile, stuck	65	95
Round, stuck	66	95
Unfired round or misfire	64	95
Replenisher assembly, 90-mm gun	9	14
Reset safety, loader's	24	30
Reticle:		
Lamps, rangefinder	29	32
Pattern:		
Periscope	30	40
Rangefinder	29	32
Rheostat, rangefinder light switch	29	32
Safety:		
Loader's:		
Reset	24	30
Traverse	22	29
Precautions	61	93
Scale transfer lever, rangefinder	29	32
Scales, rangefinder	29	32
Scope of:		
Manual	2	3
Tank gunnery qualification course	99	148
Scores, gunner's preliminary examination	106	156
Secure guns (drill)	70	101
Sensings	92	137
Moving target	94	138
Service:		
Firing exercises	108	157
Piece, of the	43, 59	83, 93
Shifting mechanism, traversing mechanism	40	57
Short sensing	92	137
Shot adjustment exercises	107, 108	156, 157
Sight adjustment examination	104	151
Sights	28	32
Simulated firing examination	105	153
Slip clutch, traversing mechanism	40	57
Solenoid, machinegun	13	19
Spare parts, stowage	88	123
Special conditions, direct fire in	98	148
Stationary targets, subsequent fire command	93	137
Stereoscopic:		
Pattern, rangefinder	29	32
Vision, training	33	47
Stowage	43	83
Ammunition	72	103
List	88	123
Stuck:		
Projectile, removing	65	95
Round, removing	66	95
Subcaliber firing exercises	107	156
Subject schedule for training	app II	181

	Paragraph	Page
Subsequent fire commands	93	137
Moving target	94	138
Supercharge pump	41	66
Superelevation transmitter	29, 41	32, 66
Switch:		
Coaxial machinegun	23	29
Light, rangefinder	29	32
Master relay	25	31
90-mm gun	23	39
Traverse safety	22	29
Turret motor	23	29
Synchronization of linkage	34	50
Tables, tank gunnery qualification course	107, 108	156, 157
Tank:		
Commander (see also Crew members):		
Power control. (See Control.)		
Gunnery qualification course	99	148
Target (see also Moving and Stationary):		
Sensing	92	137
Techniques, range finder training	33	47
Terminology, use of definite	51, 52	85
Testing:		
Azimuth indicator	36	51
Gunner's quadrant	37	55
Tools, stowage	88	123
Tracking	94	138
Training:		
Maintenance and inspections	81	111
Range finder	33	47
Subject schedules	app II	181
Traveling lock, gun	26	31
Traverse:		
No-back	21, 40	29, 57
Safety, loader's	22	29
Traversing controls. (See Controls.)		
Triggers, firing	24	30
Tripod mount, machinegun	16	26
Tube fittings, flareless	42	82
Turret:		
Controls. (See Controls.)		
Commands for	52	85
Examination	104	151
Lock	26	31
Motor switch	23	29
Mounted machinegun. (See Machinegun.)		
Power operation	27	31
Unfired round, unloading	64	95
Unit of error	33	47
Unloading:		
Stuck:		
Projectile	65	95
Round	66	95
Unfired round, misfire	64	95

	Paragraph	Page
Use of:		
Ammunition	72	103
Definite terminology	51	85
Valves, power controls	41	66
Vision devices	28, 35	32, 51
Wounded, evacuating from tanks	77	109
Zeroing:		
Exercise	108	157
Machinegun	107	156
Sights, 90-mm gun	32	46

[AG 470.8 (31 Jan 55)]

By order of the Secretary of the Army:

M. B. RIDGWAY,
General, United States Army,
Chief of Staff.

Official:
JOHN A. KLEIN,
Major General, United States Army,
The Adjutant General.

Distribution:
Active Army:

Tec Svc (1)	Abn Div (5)
Tec Svc Bd (2)	Brig (2)
CONARC (40)	Inf Regt (2)
Army AA Comd (2)	Armd Gp (5)
OS Maj Comd (5)	Armd Bn (5)
OS Base Comd (2)	Armd Co (5)
Log Comd (2)	USMA (5)
MDW (1)	Inf Sch (50)
Armies (10)	Armd Sch (50)
Corps (5)	PMST ROTC units (2)
Inf Div (5)	Mil Dist (2)
Armd Div (25)	

NG: Same as Active Army except allowance is one copy to ea unit.
USAR: Unless otherwise noted, distribution applies to ConUS and overseas.
For explanation of abbreviations used, see SR 320–50–1.

U. S. GOVERNMENT PRINTING OFFICE:1955 O— 334854

VICTORY AT SEA

ALL 26 EPISODES   3 BLU-RAY DISCS

PERISCOPEFILM.COM

VICTORY AT SEA

HI-DEFINITION BLU-RAY 1080/24P

BLU-RAY 1080/24P

VICTORY AT SEA

VICTORY AT SEA

VICTORY AT SEA

DISC 1 OF 3

Blu-ray

IN HIGH DEFINITION
NOW AVAILABLE!

NORTH AMERICAN P-51 MUSTANG
PILOT'S FLIGHT OPERATING
INSTRUCTIONS

P-51 MUSTANG

RESTRICTED

COMPLETE
LINE OF
WWII
AIRCRAFT
FLIGHT
MANUALS

WWW.PERISCOPEFILM.COM

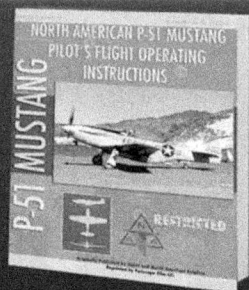

©2013 Periscope Film LLC
All Rights Reserved
ISBN#978-1-940453-01-9
www.PeriscopeFilm.com

www.ingramcontent.com/pod-product-compliance
Lightning Source LLC
Chambersburg PA
CBHW051959090426
42741CB00008B/1467

* 9 7 8 1 9 4 0 4 5 3 0 1 9 *